BETWEEN TWO SEAS

Between Two Seas

THE CREATION OF THE SUEZ CANAL

Lord Kinross

JOHN MURRAY

© Lord Kinross 1968

Printed in Great Britain for
John Murray, Albemarle Street, London
by The Camelot Press Ltd, London
and Southampton
7195 1813 x

To
ILLTYD HARRINGTON

Contents

Illustrations

ILLUSTRATIONS

† Compagnie Financière de Suez, Paris
‡ Radio Times Hulton Picture Library
* The Illustrated London News

Foreword

This book coincides with the hundredth anniversary of the opening of the Suez Canal, which falls in 1969. It covers the story of its creation and of the political and personal forces involved in it. This international waterway, which changed the world's geography and helped to change its history, was no product of international agreement. On the contrary, it was achieved in defiance of international dissension and opposition, largely through the vision and pertinacity of a single man. It was a product of nineteenth-century individualism, triumphing over the doubts and suspicions and animosities of nineteenth-century nations.

The creator of the Suez Canal was a Frenchman, Ferdinand de Lesseps. The nation which steadfastly opposed its creation was Britain. The reason for her opposition was fear of France, of a French predominance in Egypt which would upset the strategic balance of the Mediterranean and endanger the British position in India. This attitude discouraged France from affording the Canal more than qualified official support.

The result was a long spell of diplomatic warfare, fraught with abortive negotiation and obstructive intrigue. This political conflict proved more formidable than any technical or financial problems as an obstacle to the creation of the Suez Canal, which thus took fifteen years to complete. Its completion was a victory against British odds for French enterprise and for a Frenchman's persistence.

Its opening was followed by an abrupt and ironic reversal of British policy. Inevitably Britain emerged as the chief beneficiary of the Canal which she had so consistently opposed. It became an essential link in the chain of her communications East of Suez, and thus a major British national interest. To safeguard this for the future, Britain took two steps. Firstly, to ensure a voice in the Canal's affairs, she acquired a substantial shareholding in the Suez

Canal Company, in which she had previously refused to invest. Secondly, to ensure the Canal's protection, she herself occupied Egypt, as she had once feared that France would do, and thus embarked, at the expense of French aspirations, on a new policy of landward commitments on the African continent.

Such is the historical prelude to the events of the present century in relation to the Suez Canal, and to the complex problems now confronting it at the start of its hundredth year.

Acknowledgments

For documentation on the origins of the Suez Canal, I have drawn extensively on two unpublished sources. These are, firstly, the British Foreign Office records, with especial reference to the Suez Canal series, in the Public Record Office, London, transcripts of which, under the terms of Crown Copyright, appear by permission of the Controller of Her Majesty's Stationery Office; secondly, the diplomatic archives of the French Ministry of Foreign Affairs, Quai d'Orsay, Paris, especially the series *Egypte, Correspondance Politique*, Vols. 25–46; *Turquie, Correspondance Politique*, Vols. 333–363; and *Mémoires et Documents, Egypte,* Vols. XIII and XIV.

I am grateful besides to Brigadier H. Bulwer-Long for access to the papers of Sir Henry Lytton Bulwer (Lord Dalling and Bulwer), and to the National Register of Archives for help in this connection.

With regard to general published sources, I have relied especially on the five volumes of Ferdinand de Lesseps's *Lettres, Journal, et Documents pour servir a l'Histoire du Canal de Suez*; on the two volumes of his biography, *Ferdinand de Lesseps*, by Georges Edgar-Bonnet, written with access to the private archives of the Suez Canal Company and of the Lesseps family; on the two volumes of *L'Isthme et le Canal de Suez*, by J. Charles-Roux; and on *The Suez Canal: Its History and Diplomatic Importance*, by Charles W. Hallberg, Ph.D., of Syracuse University, U.S.A.

Since this is a book designed for the general reader, I have refrained from encumbering its pages with detailed source references, in footnotes or otherwise. Instead I have listed these references in more general terms, chapter by chapter, in a Bibliography at the end of the book.

I am indebted to the Egyptian Suez Canal Authority, in Cairo and Suez, for facilities in visiting the Canal; to Mrs. Joan St.

George Saunders, of Writer's and Speaker's Research, for researches largely into newspaper and Parliamentary sources; to Mr. C. E. Stevens, of Magdalen College, Oxford, for reading the typescript and proofs; to Sir Roy Harrod, for vetting my arithmetic; and to Miss Diana de Zouche, for typing an often barely decipherable manuscript.

CHAPTER I

Napoleon at Suez

From one sea to the other, between the two continents, the Suez Canal strikes broad and straight. Northwards lie the cool waters of the Mediterranean, the Atlantic Ocean; southwards the warmer waters of the Red Sea, the Indian Ocean. Contrasting landscapes of Asia and Africa mark its course: beyond the East bank the arid desert of Sinai, beyond the West abundant fields and palm-groves fed from the valley of the Nile. Moving between them, in a measured majestic procession, the ships of the world link the hemispheres. In cutting a canal through the Isthmus of Suez, just a century ago, engineers first materially changed the geography of the actual world, as explorers before them had changed that of the known world—both with consequent changes in its economic and political balance.

Above Suez, a shallow reach of marshy ground winds in through the cultivation towards the West bank of the Canal. Here at the turn of the year 1798—in a landscape still a desert, with no cultivation—Napoleon Bonaparte stood. He had found what he was looking for—the bed of the ancient canal of the Pharaohs, disused for a thousand years. Napoleon had invaded and occupied Egypt that summer and was now in possession of Cairo. The campaign had been planned, as an alternative to the more speculative invasion of Britain, to strike at his enemy not at home but through her Eastern Empire. 'Really to destroy England', he had declared, 'we must get possession of Egypt.' For it commanded the Mediterranean and the passage to India.

Thus he carried from the Directory an order to the Army of the East. It must 'drive the English from all their Eastern possessions, wherever it goes, and in particular it will destroy all the trading stations of the Red Sea'. To open up a new route for

French trade to the East, he was also instructed 'to arrange for the cutting of the Isthmus of Suez'—in short, to unite the two seas.

Thus on Christmas Eve Napoleon started out across the desert on a three-stage journey from Cairo to Suez. He followed the track used by pilgrims to Mecca and traders from India. He was accompanied not only by senior staff officers but by his Chief Engineer and other members of his *Institut d'Egypte*. Modelled on the *Institut de France*, this was a learned team of scholars and technical experts. It was to prepare a survey of the country in its manifold aspects, for the benefit not only of Egypt itself but of France and all mankind.

This illustrious company travelled in some discomfort. The nights were cold, and the gravel wastes were almost devoid of fuel. Marking the pilgrim station of Haura was a prodigy of the desert, a single aged acacia tree, sacred to the one God and festooned with rags from the clothing of the Faithful. Napoleon took his Christmas sleep beneath it, lest his soldiers cut the branches for firewood. Instead they made what fires they could from camel-thorn and the bones of dead animals. They carried loaves of bread on the points of their lances, and skins full of water around their necks. For Napoleon himself there were also some chickens.

A six-horse carriage accompanied him, but he preferred to ride, greatly impressing by his hardiness and by the simplicity of his equipage a group of Cairo merchants. Products of a decadent Mamluk régime, they had joined his caravan, with retinues of personal servants, to take advantage of the unaccustomed security. On his arrival in Suez, Napoleon refused the offer of a house, and remained in his tent. The houses were in any event barely habitable. Suez was a broken-down port, 'a squalid and filthy place', with a silted-up harbour, derelict shipyards, and only a few Arab dhows in the roadstead.

But Napoleon at once saw it as a suitable bridgehead for his planned conquest, in the footsteps of Alexander the Great, of Syria, Mesopotamia, Persia and India. He gave instructions accordingly for the harbour to be deepened and, as the nucleus of a naval establishment, for local craft to be repaired, boats to be sent from Cairo, and larger keels to be laid down in the Nile shipyards. He interviewed an Indian, who claimed to be an agent of Tippoo Sahib, the pro-French Sultan of Mysore and a likely ally, within the enemy's gates, for the conquest of India. Suspicious of the man's credentials, Napoleon wrote to Tippoo in person, promising to deliver him from the British yoke with his great and invincible army, and asking him to send an accredited envoy to Cairo.

Napoleon interviewed sea captains and merchants from various parts of the Arabian peninsula, gaining intelligence about the state of their countries. To conciliate the Arabs, he granted protection to the Beduin of Sinai, as he had already done to the Christian monks of its monastery, that they might 'transmit to future races the tradition of our conquest'. To pay respect to the Jews and their Biblical connection with Sinai, and at the same time to make sure of his water supplies, he crossed to the peninsula to spend a day by the Wells of Moses. With his officers he drank from them as Moses had done, finding their water to be brackish but drinkable, and ordering their occupation as a watering station for his forces in Suez. Meanwhile Gaspard Monge, the distinguished savant who directed his Institute, enlightened the company with a discourse, quoting suitable references from *Exodus,* on the flight of the Israelites across this Red Sea, and their subsequent wanderings in the Sinai desert.

On their way back to Suez in the evening, Napoleon and his group almost met with the fate of the Pharaoh and his hosts, pursuing the Israelites. The tide began to rise over the ford, at the head of the Gulf, where they crossed. It grew dark and the native guides, whom the soldiers had primed with eau-de-vie, lost their footing on the terra firma of the sandbank. The sea rose to the bellies of the horses, which started to swim. Some stumbled into the water. There were cries for help, and Napoleon exclaimed, 'Did we come here to perish like Pharaoh?' One of his generals, the engineer Caffarelli, who had lost a leg in the German campaign and was nicknamed 'Old Wooden-leg', mislaid the artificial limb and was on the verge of drowning when he was rescued by a stalwart N.C.O. Old soldiers who knew their

Bible recited appropriate passages from the Old Testament.
But those on shore demolished a house, for want of fuel, and
set fire to it as a beacon to guide the floundering troops. Thus
all reached land in safety. The rescuer of Caffarelli was promoted
and invested by Napoleon with a sword of honour, grandly
inscribed on one side with a citation of his heroism, and on the
other with the historic words 'The Crossing of the Red Sea'.

But Napoleon's main concern, after providing for the port of
Suez, was to pierce its Isthmus. Having located the bed of the
ancient canal, he and his experts proceeded northwards along it
for fifteen miles towards the Bitter Lakes—the long shallow
lagoon which interrupts the present ship canal, and from which
the original canal ran down, by a parallel course, to the Gulf of
Suez. They found its traces, with banks well marked, and a
depth sufficient to conceal a horseman from sight. Turning away
westwards towards Bilbeis and the ruins of Bubastis (now
Zagazig), they saw signs of another stretch of it, apparently
linking the lakes with the former Pelusiac branch of the Nile.

Napoleon had seen enough to convince him that a new canal
could be dug, perhaps following part of the line of the old. After
a few skirmishes in the desert with Beduin tribesmen, his caravan,
enriched with their horses and camels, returned to Cairo. Here
he immediately instructed Jean-Baptiste Le Père, his Chief
Engineer, who had accompanied him, to carry out a complete
survey of the Isthmus of Suez, establishing its levels with a view
to cutting a canal from one sea to the other.

*   *   *

It seems certain that the Isthmus of Suez was once not an
isthmus but a strait. At a relatively late geological period the
Mediterranean was still joined to the Red Sea by a shallow
channel, which linked a chain of lakes. Then gradually the two
seas thrust out deposits of soil, southwards and northwards,
until a neck of land divided them. This process of encroachment
continued right into historical times, leaving no channel but
only the lakes and a depression between them. It was matched,

4

moreover, by the gradual silting up of the Nile Delta and the formation of the lands of Lower Egypt. The Pharaohs thus had to face the perennial problem of keeping communications open, with seas receding and river channels diminishing as the centuries went by.

Hence from earliest times the Egyptians became proficient in canalization. The canal discovered by Napoleon was ascribed by such classical authors as Strabo and Pliny to Sesostris, a monarch of semi-legendary achievements in the second millennium B.C. In fact in his time, and for many centuries to come, little more was done than to maintain existing natural channels. This artificial canal, from the Nile to the Bitter Lakes, and from there to the Red Sea, was more certainly the work of Necho, who reigned in the seventh century B.C. So writes Herodotus, adding that 120,000 Egyptians perished in digging it, that it was four days' journey in length, and wide enough for two armies abreast. At this time the Red Sea still penetrated some thirty miles into the Isthmus beyond the present Gulf of Suez, probably as far as the Bitter Lakes themselves. But it was slowly receding, and it thus became necessary for the forces of man to offset and supplement those of nature.

Necho's canal was continued by Darius, the Persian conqueror, wishing to improve his communications with the Persian Gulf. But it was left unfinished on account of a traditional belief that the Red Sea was higher than the Mediterranean, and could thus inundate Egypt or pollute its fresh water. The early Ptolemies none the less finished it and introduced sluices—a canal fifty yards wide and deep enough for large vessels; but their successors neglected it and Cleopatra, bent on flight to the Indies after the Battle of Actium by transporting her fleet across the Isthmus, found it unusable. The Emperor Trajan revived it. But under the Christians it fell again into disuse.

The Arab invaders of the seventh century restored the canal once more, on the instructions of the Caliph Omar, 'the Prince of the Faithful', as a means of supplying Medina and Mecca with the abundant produce of Egypt. But a subsequent Caliph filled it in, during the latter part of the eighth century, to deprive Medina, by then in revolt, of these resources. Thus the man-made waterway which had existed in some form or another, now flourishing, now languishing, for fifteen hundred years—the Canal of the Pharaohs, the Ptolemies, the Caliphs—vanished from history for a further millennium.

Compared to the Suez Canal which Napoleon envisaged, it had always a limited scope. Its main purpose was to serve the local needs of Egypt, to keep the Delta of the Nile in communication with the Gulf of Suez and the lands adjoining the Red Sea. In any event, as time went by, it came to be seen by the Egyptians more as a danger than as a convenience. For it was the artery that led to Palestine and Syria, hence a channel not merely for commerce but for potential invasion by the foreigner. The Isthmus of Suez became a frontier as much as a trade route and, partially in a defensive spirit, the Pelusiac and adjoining branches of the Nile were, one by one, allowed to dry up, leaving it only those two mouths—at Damietta and Rosetta—which survive today.

Napoleon, as befitted a European invader with dreams of world conquest, saw the Isthmus of Suez in a global perspective— as a means of establishing the military and political power of France East of Suez, at the expense of her traditional enemy, Britain. Moreover France, in a mercantile spirit, sought to challenge British supremacy on the trade routes of the East. By opening up and controlling a new route from the Mediterranean to the Indian Ocean, she hoped to regain a share of that oriental trade of which a momentous event had deprived her three centuries earlier.

At the end of the fifteenth century the Portuguese explorer Vasco da Gama, navigating with an unfamiliar instrument, the compass, had discovered the Cape of Good Hope. His discovery, and subsequent voyages, opened up a great new route to the East, which radically changed the world balance of trade. It started an era of long sea voyages by ocean-going vessels, which by-passed the Mediterranean and the Red Sea and assisted the rise of new mercantile nations—the Portuguese, the British, the Dutch.

In the Mediterranean it led to the ruin of the maritime Republic of Venice, which had grown into a rich commercial power and might well, in the opinion of Voltaire, have come to dominate

6

Europe. But the Venetians could not rise to the challenge of the Atlantic, beyond these sheltered waters. In alarm they tried at once to promote a canal through the Isthmus of Suez—the first attempt for seven hundred years to revive this ancient route for shipping. But their scheme was defeated by the Portuguese, rounding the Cape and moving northwards to control the Red Sea. Thus support for a canal was not forthcoming.

The French too were hit by the change, if to a lesser degree than the Venetians. Eventually they were able to rise in part to the challenge, competing with their rivals from their own Atlantic ports. Meanwhile however Marseilles, with the other ports in the Mediterranean, suffered severely from the diversion of the oriental trade. This had been a prosperous interest of France, as of Venice, since the time of the Crusades. The thirteenth and fourteenth centuries had seen a mercantile renaissance throughout the Mediterranean. French merchants established themselves in the former ports of the Phoenicians in the Levant, and in Alexandria, where they built up entrepôt markets for the exchange of the products of Europe with those of the Far East, India and Arabia. Merchandise was conveyed through Arab agents over the Asiatic caravan routes or across the Isthmus of Suez by means which were slow and laborious but still unchallenged by competition.

The trading stations of the Levant disappeared with the loss of the Crusader kingdom of Jerusalem. That of Alexandria continued to flourish, but was now in its turn condemned by the circumnavigation of Africa. A new world had been born, and a new mercantile conflict, between the Axis of the Atlantic and that of the Mediterranean; and this was to continue for four hundred years.

The French did their best, through the Ottoman Turks—now masters of the lands of the Eastern Mediterranean—to save what they could of their eastern trade. In the sixteenth century François 1er concluded a treaty with the Sultan, receiving privileges in the Levant, which at least favoured French commerce with Syria and the Euphrates valley. The Turks later conceived the idea of developing and extending this route, both by land and by sea, to the Indies through Baghdad, Basra and the Persian Gulf. Nothing came of this. But in European hands the plan for a Euphrates route was to alternate at intervals through the following centuries with the plan for a route through Suez.

*　　　*　　　*

The dream of joining the two seas remained alive meanwhile, if only in the minds of individual sages or scientists, after the failure of Venice to achieve it. Towards the end of the sixteenth century it was revived by a renegade Calabrian adventurer, Haji Ali, who, after spending fourteen years of his youth as a galley-slave in Algiers, turned Moslem, acquired a ship of his own, campaigned against the Christians in the Mediterranean, and rose to be Admiral-in-Chief and a man of power at the Court of the Sultan of Turkey. Fighting the Portuguese in the Red Sea, with a fleet which he had brought in sections overland, he persuaded his sovereign to re-establish, for its convenience, the ancient canal. Turkey, however, became involved in an expensive war with Persia, and the plan was shelved. Richelieu was approached half a century later with a similar proposal. But by now the French merchants, thanks to the formation by Louis XIV of the *Compagnie des Indes,* with monopolistic trading rights in the Atlantic port of Lorient, were moving over into competition on the Cape route, hence had momentarily lost interest in Suez.

Another move to reopen the route through Suez came from the seventeenth-century German philosopher Leibnitz. In a memorandum to 'the very Christian King' Louis XIV, he urged for this purpose nothing less than a Crusade for the conquest of Egypt and the destruction of the Ottoman Empire. He received only the dry response from the French Secretary of State that the idea of a Holy War had 'ceased to be in fashion since St. Louis'. A little earlier, in the sixteenth century, it had indeed met with little support, when Pope Sixtus V fathered the idea of a European Christian league against the Turks, opening up the Isthmus for the purpose not of trade but of transporting Catholic missionaries to India.

A little later, in the early eighteenth century, it was to be revived in a more mercantile but no more acceptable spirit by another philosopher, with economic leanings, the Marquis d'Argenson. A man so Utopian as to merit, in the opinion of Voltaire, a ministry in the Republic of Plato, he envisaged a grand Crusade with prodigious commercial advantages to the Christian powers, who would share between them a canal between the Red Sea and the Mediterranean, in unity and to the exclusion of Islam.

Among the abortive canal schemes of these centuries, only one deserved to rank as a serious proposition in practical terms. This was embodied in a late seventeenth-century work by Jacques Savary, a distinguished merchant and economist, which set forth in precise detail various alternative routes for a Suez Canal. He took proper account of the financial and technical problems involved, but concluded that the plan could only materialize were his sovereign first to become master of Egypt. Savary, however, was too far ahead of his time, technically and politically, to attract the serious attention of Versailles.

None the less, in the course of the reign of Louis XIV, the establishment of a trade route across the Isthmus of Suez, whether overland or by canal, became for the first time an objective of French national policy. French traders still remained in Egypt, braving, now alone among Europeans, the insults and extortions of the Turkish authorities. Their tenacity attracted the notice of Colbert, who came to see the Isthmus of Suez as 'the gateway of communication between two different worlds', hence of a route which should be reopened to the exclusive advantage of France. Louis XIV accepted this view, which at first envisaged the creation of a *Compagnie du Levant,* with a monopoly parallel to that of the *Compagnie des Indes* in the Eastern Mediterranean. Representations were thus made in 1665 to the Sublime Porte, the seat of the Ottoman Government in Constantinople, requesting the right to establish warehouses at Suez and a guarantee of security for the transport of merchandise thence, whether overland or down the Nile, to Alexandria.

The French argued that the reopening of the route, with its substantial Customs revenues, would greatly benefit not only themselves but the Ottoman Empire, helping to restore that paramount position as an avenue of trade between East and

West which it had lost through the discovery of the Cape. But the Turks were ill-disposed to heed so rational an argument. Their Empire was now entering on a period of decline, to which the loss of that trade had contributed, and which bred an increasing antagonism towards the West. Fanaticism was all too apt to prevail in Ottoman counsels, and the French were reminded that the Red Sea was closed to Christian shipping north of Jedda. It was a Moslem lake and would remain so, lest infidel sailors desecrate the tomb of the Prophet Mohammed or even attempt to remove it.

But the traditional line of French policy, initiated by Colbert, persisted. It was pursued, if intermittently and without concrete results, by a series of ministers of the *ancien régime*. In 1685 the French Ambassador in Constantinople resumed negotiations with the Porte, officially proposing for the first time the construction of a canal through the Isthmus. But the Turks took refuge once again in the religious objection, now influenced also by the growing dissidence of the Mamluk beys of Egypt, against whom they were reluctant openly to assert their authority. As time went by, these turbulent vassals, with their growing awareness of the value of the Red Sea trade, were to become a key factor in the affairs of the Isthmus.

Under Louis XV attention shifted for a while from Suez to the Euphrates route, with the development of French trading stations in Baghdad and Basra. Then it shifted once more, after the end of the Seven Years' War, with its French colonial losses, to the definite project of an occupation and colonization of Egypt. The Duc de Choiseul, in 1769, judged the time to be ripe for this operation, now that the Mamluk Ali Bey had overthrown the Sultan's authority, achieving effective autonomy and promising support for a route across the Isthmus of Suez. But Choiseul was thrown out of office, and soon afterwards Ali was murdered.

Under Louis XVI a forceful protagonist of the canal was the Baron de Tott, a Hungarian instructor of the Turkish army. Sent by the French government on a mission to Egypt, he reported strongly in favour of the re-establishment of the ancient waterway of the Caliphs, voicing the pious hope that in the future its

foundations for social good might become 'as solid as the prejudices designed to destroy them'. He elicited a promise of support
from the Sultan, which did not however materialize.

Meanwhile the French Red Sea trade contrived to carry on
within the limitations imposed on it. It specialized in coffee from
Arabia and tissues from India, transported from the Red Sea
ports over the traditional caravan routes and by the Nile across
Egypt to the Mediterranean. Thus Alexandria, an entrepôt port
reduced to a mere five thousand inhabitants, still harboured a
French commercial community. But now, in the latter part of
the eighteenth century the British, who had for long left this
field, such as it was, more or less to the French, began to appear
on the scene. It was dawning on their merchants in India that
Suez offered a passage far shorter than that of the Cape, and
moreover, if its obstacles could be overcome, more secure than
the Euphrates route, which was beset by increasing tribal disorders. Following the French example, they started to come to
terms with the Mamluk beys, who in effect now governed Egypt
in disregard of the local Turkish authorities.

*   *   *

Initially this was largely the fruit of individual enterprise. The
first British pioneer was James Bruce, a giant of a Scotsman, with
the traveller's questing spirit, a taste for archaeology and a talent
for sketching classical ruins. He reached Egypt, dressed as a
dervish, in 1768, after an adventurous spell as British Consul
among the Barbary pirates of Algiers and a shipwreck off the
African coast. To his surprise he found none of his countrymen
at work in so promising a land. After exploring Ethiopia and
discovering the source of the Blue Nile, he returned to Cairo.
Here, making up for lost British time, he established contact not
only with the Bey then in power, the successor of Ali, but with
the British authorities in India. Here he found a supporter in
Warren Hastings, then Governor of Bengal. The result was a
trading agreement between Governor and Bey, which permitted
ships to sail up to Suez with goods and despatches.

Britain's next pioneer in the field was George Baldwin, an enterprising merchant of independent outlook, who had been trading for some years from Cyprus. A man quick to see that Egypt, from its position between the two seas, afforded 'greater commercial advantages than any other land upon the globe', he began to explore the prospects of a commercial route to the Indies through the Red Sea. The Bey encouraged him, promising: 'If you bring the India ships to Suez, I will lay an aqueduct from the Nile to Suez and you shall drink of the Nile water.'

He went to Suez, where a vessel was expected from India. But after thirty days it had failed to appear. 'My provisions', he afterwards wrote, 'were exhausted; my spirits impatient; the desert barren, indeed of all resource. . . . I thought I should get to India sooner by way of England; and the holy caravan passing by Suez at this time on its return from Mecca I mounted a dromedary and accompanied them to Cairo.' Baldwin was back in Alexandria a year later as agent of the East India Company for the transport of 'immediate' mails. Patriotically—and too optimistically—he wrote of the prospects: 'We composed our bowl of the Ganges, the Thames and the Nile, and from the top of the Pyramid drank prosperity to England.' Baldwin's personal interest was still bound up with merchandise rather than mails.

But the suspicions of the Turks, whose opposition to the foreigner grew stronger as their internal situation grew weaker, had been aroused by these British activities. The agents of Britain were accused in a firman by the Sultan, of 'sliding into Egypt' to procure maps of the country and return as invaders. In this and a later firman he reaffirmed the ban on foreign ships, decreeing that 'no Frankish vessel should approach Suez, whether openly or secretly', for its Gulf was 'the privileged route of the glorious pilgrimage to Mecca', and to permit these 'children of error' to enter it would be 'to betray religion, the Sovereign of all Islam'.

Baldwin sought the support of the British Ambassador in Constantinople, but was coldly informed that it was British policy to support the sovereignty of the Sultan against the Mamluk Beys. Thus he must respect the firmans forbidding trade through the Red Sea. In vain did he contend that such a policy would play into the hands of the French, who had all too evident designs not only on the Red Sea trade but for the conquest of Egypt itself. It was a conquest which would afford them, in his view, 'the master-key to all the trading nations'.

Baldwin none the less was sent back to Egypt in a Consular post, with instructions to ensure the free passage of mails but to suppress commercial traffic. More merchant than consul, he disobeyed these instructions, and tried to import goods through Suez regardless of the Ottoman firmans. Discredited, he fled the country. Cold-shouldered by the Ambassador in Constantinople, he returned home, where he zealously lobbied officials on the dangers of French penetration.

Proved right in 1785, when the French secured a preferential trading alliance with the Beys, in defiance of the Sultan, he was sent back to Egypt, as Consul-General, with instructions to do likewise for Britain. But little came of his mission. The Turks regained some control over the Beys. The French themselves ran into difficulties. Baldwin finally secured a treaty, but by then the British Government had lost interest. His Consulate-General was abolished, and Baldwin himself retired to the grotto of the Apocalypse on the island of Patmos, consoling himself for a while with the mystical reflections and writings to which, in his leisure moments, he had always been prone.

Neither he nor Bruce, both patriots working in their country's interest, had received fair support from their Government. This neglect was due largely to the influence of the East India Company, the dominant British power in India since the seventeenth century. The Company opposed the Suez route from the start, as a source of competition with its route around the Cape. It was an opposition fraught with difficulties for the future—and indeed matched by the opposition of the *Compagnie des Indes* to the parallel designs of the French.

Now the French Revolution and the advent to power of Napoleon changed all. Talleyrand, stressing the advantages of colonial expansion as a sequel to revolution, recalled that Egypt had been a province of the Roman Republic and could as easily become a province of the French Republic. Dundas, the British War Minister, belatedly urged that 'the possession of Egypt by any independent power would be a fatal circumstance to the interests of this country'.

But Napoleon was already there. There was an end to vague aspirations and fancies for the Isthmus of Suez. The long-projected canal was now to pass from the realms of theory into those of factual enquiry. For the first time a practical survey was to be made by technicians on the spot, to settle whether or not, and if so by what means, the two seas could be joined.

*   *   *

Entrusted with this momentous topographical survey, Le Père lost little time in returning to the deserts of the Isthmus—wastes of sand, as he saw them, peopled only by ostriches and gazelles, predatory vultures and eagles, and marauding Beduin tribesmen. With a team of engineers and surveyors, which included his brother, he reached Suez in the middle of January 1799. Within a few days, escorted by a detachment of Maltese legionaries, they had started to trace Napoleon's steps along the bed of the ancient canal towards the Bitter Lakes.

It was their specified task to determine by geometrical measurement the levels of the land between the two seas—together with those of the Nile and the seas themselves. Throughout they were harassed by abnormal difficulties. They were short of instruments, which had failed to arrive from France. Their means of transport were inadequate. Hunger and thirst beset them: stores were severely rationed, and periodically a shortage of water led to an interruption of the work and a return to Suez or Cairo to replenish supplies. Treacherous or incompetent guides misled them, separating them from their escort and laying them open to the Beduin raids. Moreover, they had only been at work for three weeks when Napoleon launched his invasion of Syria. This deprived them of their escort, and obliged them to postpone for a further six months the main part of the survey.

When they resumed, it was still under war conditions, with

enemies multiplying around them. The escort would be reduced without warning, to fulfil more urgent military duties. The soldiers would leave them unprotected, as they sought encounters with the Arabs and pursued them across the desert; an Arab guide was suspected of leading them away from a well to protect the passage of his own tribesmen into Syria; extremes of heat exhausted them, and they had once to make a forced march of sixteen hours in flight from a rumoured enemy landing at Suez.

Seldom has an important topographical survey been conducted in face of such continuous hazards. The engineers surmounted them with dedicated zeal and concentration. To guard against error, which the conditions made only too likely, Le Père imposed a strict rule that if there were a grain of doubt as to the accuracy of a calculation, the ground must be covered again—and this was done. He saw it as all the more necessary in view of the unlikelihood of a second survey, to confirm the first. In this spirit his team covered the ground in three successive expeditions.

As a result of the first, they concluded that the level of the plain of Suez was only a little above that of the Red Sea. Thus it would be easy, by means of a shallow canal, to carry its waters as far as the Bitter Lakes. Here stakes were placed across the basin to mark the line of levels. On the second expedition, in the autumn of 1799, Le Père continued his survey as far as the Mediterranean, to determine the relative levels of the two seas and those of the Nile in relation to both. For this purpose he divided his team into two groups. The first, under his own direction, worked northwards to Lake Menzaleh, the wide shallow stretch of inland water, chief receptacle of two former branches of the Nile, from which the sea is only narrowly divided. He reached the Mediterranean itself, near the traces of the ancient port of Pelusium, where the former Pelusiac branch once ran into it. This task was completed to his satisfaction. His second group, briefed to establish the degree of the fall in the flow of the Nile to the Mediterranean, achieved results which were less conclusive.

On a last expedition, in October 1800, Le Père endeavoured to judge the flood of the Nile in the Wadi Tumilat, the valley running from West to East, the Biblical land of Goshen, where the ancients had cut their canal from the river to the lakes in the direction of Suez. Here his survey was confused by an exceptional flood, which broke the dykes and submerged traces of the ancient canal. But he was encouraged to note that the Nile waters inundated the whole length of the valley as far as the Pharaonic ruins of the Serapeum, just north of the Bitter Lakes.

Soon after his return to Cairo, Le Père sent an interim summary of the results of his survey to Napoleon, who was now back in Paris, leaving his army behind him after the British defeat of his fleet in the Battle of the Nile. Le Père insisted that there could be no major obstacle to the re-establishment of the waterway of the ancients. This would link the two seas, not directly, but by means of the Nile itself and of two canals—one from the river to the Bitter Lakes and the other from the Lakes to Suez. The ancients, he believed, had abandoned this route on account of the variable levels of the Nile and of the Red Sea tides, which permitted only seasonal navigation. Under more modern conditions this problem could be solved by a system of sluices. Good administration in the upkeep of the waterway would do the rest, particularly at points in the desert where it might be threatened by moving sands.

To a project of so high a political and commercial importance as the reconstruction of these two canals, the matter of financial cost was, he suggested, the least of all obstacles. What price 25 or 30 million francs if its expenditure led to the restoration of 'the riches and the trade of India by their original and natural route'? Moreover, maintenance costs could be covered by the reclamation of the uncultivated lands of the valley, and by eventual tolls and dues.

As to the navigational hazards of the Red Sea, these had in Le Père's view been exaggerated, partly for political motives and partly through the ignorance of mariners, ancient and modern. As a route for shipping it was no worse than other narrow seas, and incomparably better than the ocean itself. This was a matter for study by expert navigators, in relation to the known difficulties, to the prevailing weather, and especially to the duration of the seasonal monsoons. Meanwhile the harbour of Suez could be improved without difficulty.

On receiving this memorandum, Napoleon wrote, in February 1801, to the Tsar Paul I, whose support he was trying to enlist: 'The English are attempting a landing in Egypt. It is in the interest of all Mediterranean and Black Sea powers that Egypt remain French. The Suez Canal, which will join the Indian seas to the Mediterranean, has already been traced: it is an easy enterprise which will require little time and which will bring incalculable advantages to Russian commerce.'

But the full implications of Le Père's survey were only clear to the world with the completion of his detailed report. His reasons for the choice of the indirect route then became evident. According to his calculations, the level of the Red Sea at high tide was more than thirty feet higher than that of the Mediterranean. This effectively ruled out the cutting of any direct canal between the two seas. For the waters of the Red Sea could inundate the lands of the Delta. So did Le Père give apparent scientific sanction to that traditional belief of the ancients which had deterred Darius from completing his canal. For all his scrupulous care, he had made a miscalculation of major proportions.

Some valid objections were raised to his estimate. Two reputable mathematicians, Laplace and Fourier—one of Le Père's colleagues in Egypt—argued that any such disparity of water levels was contrary to nature's laws of equilibrium. No one however was now likely to pay them much heed. For Napoleon's Egyptian expedition had ended in failure, after a mere two years. The British landed at Alexandria in 1801 (incidentally with George Baldwin attached to their forces). They defeated Napoleon's abandoned army and arranged for its transport back to France. By the time of the delivery of Le Père's final report, the French were no longer in Egypt. Nor were the British who, having ejected them, had no desire to remain there. Le Père himself returned to his original vocation, which was that of an architect, reconstructing Malmaison for Napoleon, erecting the column in the Place Vendôme, and eventually placing his master's statue on top of it.

The project for a Suez Canal lapsed into a period of oblivion. Thanks to his mathematical error, a generation was to elapse before it emerged once more into the light of day as a proposition worthy of serious practical notice.

CHAPTER II

The Overland Route

As the development of the compass led to the discovery of the route round the Cape, so in time did the development of steam lead to that of the route across the Isthmus of Suez. The initial impulse to the use of steamships for ocean-going trade came from the citizens and governments of British India, still thinking primarily in terms of the Cape route. 'Steam Committees' and 'Steam Funds' were established in several provinces. In 1823 the Calcutta Committee, with the support of the Begnal Government, offered a substantial prize for the first communication hy steamship between Britain and Bengal. It had been its original intention to sponsor a regular route through the Mediterranean and the Red Sea, by means of two steamships of the East India Company, one on either side of the Isthmus of Suez. But the Company, conservative and tight-fisted, ignored this suggestion, and indeed another decade was to pass before it would sanction expenditure on steam lines at all. Thus the prize came to apply to the Cape route.

In 1825, following up its announcement, a naval officer, Captain James Henry Johnston, sailed into Calcutta from Falmouth, amid scenes of some excitement, aboard a 500-ton paddle-steamer, the *Enterprise*. Pioneer though he was, his voyage was officially regarded as a failure, since it had occupied 113 days, instead of the two months anticipated. Impeded not only by head-winds and storms but by the great weight of coal which she carried, the *Enterprise* had spent only 62 days under steam, and 40 days under sail. Captain Johnston, however, was granted half the promised award. He decided to sail the oceans no longer, and devoted the rest of his career to the simpler task of developing inland steam routes over the rivers of India.

In the Hooghly, when the *Enterprise* arrived, was a young naval

c

pilot named Thomas Waghorn. Her performance confirmed a
faith in the prospects of steam upon which he had been pondering
for some time past. Rising to the challenge, he determined to
carry on from the point at which Johnston had left off. Thinking
at present in terms of the Cape route, he was ultimately destined
to become an effective pioneer of the route via Suez.

An adventurous, free-lancing Englishman, of the stamp of
Baldwin and Bruce, Waghorn came of landed stock. Born at
Chatham in the first months of the nineteenth century, he grew
up so tall that he was once excluded from a side-show at a country
fair, the showman protesting to his companion: 'I pray, Sir, take
that gentleman away. The fact is, he is two inches taller than my
giant.' Waghorn entered the Royal Navy as a midshipman, but
was paid off at the end of the Napoleonic Wars, when the strength
of the Services was reduced to a peace-time establishment.

Waghorn then signed on as third mate in a free-trading merchant ship bound for Calcutta. Here he became a pilot on the
Hooghly River, with the Bengal Marine Pilot Service. Seconded
for service with the East India Company in the Burmese War,
he commanded a cutter, with a flotilla of gun-boats, and had his
baptism of fire. It was in the course of this war that he first saw
steam power in action, studying and becoming impressed with
the work of a steam transport, the *Diana*.

He now came to believe passionately in steam but not, at this
present stage in its development, for vessels so large as the
Enterprise. What he advocated was smaller steam packets, 'buoyant
fast vessels', for the transport not of cargo or passengers but only
of mails. All space below decks would he reserved for the
storage of fuel—of enough coal, so he hoped, to last for as long
as 50 days. With recourse to coaling stations *en route,* the voyage
could be completed in 70 days—40 days less than the normal
time taken by sailing-ships. Armed with introductions and other
credentials, Waghorn proceeded to London.

Here he intended to build a ship of 200 tons, with two engines
each of 25 horse-power, to inaugurate a 'Dispatch Mail Packet'
from Falmouth to Madras and Calcutta, carrying official and
private mails. On arrival he wrote in 1827 to the office of the

Postmaster General, requesting facilities and remuneration for the vessel's expenses 'in the accomplishing of so National an undertaking'. The Secretary-General advised the Minister: 'We can have nothing to do with the establishment of steam vessels to India, and consequently cannot patronize the undertaking of this gentleman. It appears to be a wild scheme to fit a vessel for so long a voyage with Engines of the Power of 50 Horses only, which is not more than we employ upon the Voyage between Dover and Calais.' Waghorn was thus referred to the East India Company, which was responsible for conveying the mails and thus for deciding on any such proposal.

The Company gave Waghorn a lukewarm reception and, with the failure of the *Enterprise* fresh in its mind, refused to help finance his scheme. But they agreed to lend him two engines. After canvassing support from the mercantile committees of London, Liverpool, Manchester, Glasgow, and Birmingham, he returned to India. To finance his run, he set up a Steam Navigation Fund, and received promises of funds from Calcutta and Madras, but not from Bombay. Back in London once more, in search of further support, he was faced by the problem that the official rates of postage to India were laid down by Act of Parliament at twopence per letter. This was an unforeseen setback, since he would need to charge up to five shillings for what was after all a special express service. Nor, for all his persistence, was the Post Office disposed to recommend a change in the law at the request of 'this person', who appeared to a mistrustful Chief Secretary to be 'quite an enthusiast'.

Meanwhile Waghorn waited in London for the funds that the Indian Steam Committees had promised him. Here he heard the news that the Bombay Government was about to try out the *Enterprise* once more, on a voyage to Suez and back. This surely was his chance. Switching his attention immediately to the prospects of a mail route via the Red Sea instead of the Cape, he resolved to join the *Enterprise* and return to India on board her, with a consignment of mails. The Post Office, barred from entrusting letters to an individual, refused to co-operate. But the East India Company finally yielded to his importunities, and decided to give him a trial. Thus in October 1829, with a courier's passport, mails from the Company for Sir John Malcolm, the Governor of Bombay, and instructions to make 'a plain sailor-like report' of his journey, Waghorn left for Egypt, crossing Europe via Paris and Trieste.

Knowing only a few words of Hindustani and Arabic, apart from his native tongue, he took with him a distant relative, a boy of sixteen, who knew a few words of French and Italian. A young man in a hurry, bent on speed at all cost for the sake of his mails, he was frustrated in his time-table by abnormal hazards. The Simplon Pass was closed by avalanches and broken bridges and he had to take the Mont Cenis route, which was 70 miles longer. In Italy he learned that the steamer service from Venice to Trieste was suspended, and he had to proceed by the roundabout route across the head of the Adriatic. These detours added 130 miles to his journey as planned. Nevertheless he reached Trieste in nine and a half days from London, a 'record' which startled the Foreign Office, well aware that the normal Post Office service took fourteen days.

At Trieste Waghorn met with further delays. The ship for Alexandria had sailed the evening before, and he was unable to catch up with her despite a race down the Istrian peninsula. Another, loading cargo, would not sail for three days. But he bribed the captain to sail in two. After sixteen weary days he reached Alexandria, to find that the Consul-General, Barker, the agent of the East India Company, had left for his country quarters at Rosetta, by the Western mouth of the Nile. He pursued him there, by donkey and by a boat down the river, navigating it himself to take soundings and collect information which might become useful. Barker, an old hand, who knew his East India Company, warned the eager young man: 'Don't be too sure of success. I wrote about this a long time ago, but the Court of Directors *won't have it.*' From Rosetta, Waghorn sailed upstream for Cairo in a large felucca. But it grounded on a shoal and he completed his journey by donkey. Travelling across the desert and taking notes on its various features, he reached Suez—only to find that the *Enterprise* had not yet arrived.

Too impatient to wait, he left after two days, braving the ill-famed reefs and squalls of the Red Sea in an open boat, without compass or chart and with a mutinous crew, in the hope of intercepting her *en route*. Finally, without doing so, he reached Jedda, having covered six hundred-odd miles in six and a half days. Here he learned that the *Enterprise,* held up by defective engines, had never left Bombay.

He was stricken with a high fever, and was only able to continue his journey to India six weeks later. He sailed aboard the *Thetis*, a naval vessel from the Bombay Marine, which was engaged on a survey in the Red Sea. He finally reached India four months and 21 days after his departure from England. This represented 84 days' travelling time, for a journey which he had hoped to complete in 55.

On board the *Thetis* Waghorn encountered a rival in another young Englishman named James W. Taylor. Brought up in India, Taylor had likewise been trying for some years to establish a steam route, first via the Cape but now via Suez, but with passengers rather than mails. He had left London a week before Waghorn, and had reached Alexandria more quickly via Marseilles and Malta. Having issued a prospectus of his scheme in London, and obtained some support, as Waghorn had, from Calcutta and Madras, he now sought to obtain it from Bombay. Here he had little success. The Governor, Sir John Malcolm, had himself launched a scheme for a monthly service to Suez of steamers manned and financed by the Bombay Government. Indeed, following the breakdown of the *Enterprise,* it had just been inaugurated by the departure of another steamer—the *Hugh Lindsay*, diplomatically named after the chairman of the East India Company, whose reactions to so extravagant a venture were dubious.

Thus Taylor switched to the alternative scheme which he had kept in reserve—the development of the Euphrates route, from the Persian Gulf. Entrusted with despatches, he left within a few weeks for England via Mesopotamia, where his brother was Consul in Baghdad and Basra. He surveyed reaches of the Euphrates River for steam traffic, obtained exclusive navigation rights from the Pasha of Baghdad, and decided that the route was 'in every respect preferable to that of the Red Sea'. Unfortunately, on his way home, he and his party were attacked by a large force of Arabs, bent on plunder. Instead of prudently submitting, as more experienced travellers might have done, they were impelled by an excess of gallantry to resist, and Taylor, with two of his companions, was killed. Such was the sad end of the first attempt to develop the Euphrates route by steam.

Waghorn, his rival, survived to become a conspicuous figure in the history of the Isthmus of Suez. His chequered attempt at a trial mail run by the Suez route could hardly be reckoned a success. He was not however a young man to be easily daunted, whether by the forces of bureaucracy or the forces of nature. It was still his fixed idea to establish an independent steam route. Once again he received promises of support from Calcutta, whose Steam Committee conditionally placed a fund at his disposal to be applied to the construction of a suitable vessel. Once again he returned to London, where he tried to raise further capital, without apparent success. Once again he joined battle with the bureaucrats.

A plea on behalf of his Overland Mail to the Colonial Office was referred through the Treasury to the Post Office, which gave a negative reply, as before. From the East India Company he expected a better reward for the efforts he had made and the risks he had taken on his pioneer journey. In an eloquent appeal he stressed:

The vast importance of three months earlier information to His Majesty's Government and to the Honourable Company, whether relative to a war, or a peace, to abundant or short crops; to the sickness or convalescence of a Colony or district, and often times even of an individual: the advantage to a merchant by enabling him to regulate his supplies and orders according to circumstances and demands: the anxieties of the thousands of my countrymen in India for accounts, and further accounts of their parents, children and friends at home: the corresponding anxieties of those relatives and friends in this country: in a word, the speediest possible transit of letters to the tens of thousands who at all times in solicitude await them.

But in reply he was informed by the new Chairman of the Court of the Company that 'India required no steam in the East'. Nor, it seemed, were the views of the Governor-General and people of India any concern of theirs. Waghorn was advised, with a hint of menace, to return to his job in the Company's pilot service. On this he immediately resigned from the service, declaring that he would establish the Overland Route on his own and in spite of the East India Company. They accepted his resignation, taking the precaution to confirm that he would on no account be employed by the Company in the promotion of steam navigation to India.

Thus when Waghorn, two years later, proceeded on his own account to Egypt, it was (as he afterwards wrote) 'not only without official recommendation but with a sort of stigma on my sanity'. The British naval authorities seemed to believe that the Red Sea was not navigable by steam—just as, until a few years earlier, it had been thought not to be navigable in safety by sail; moreover, 'the East India Company's naval officers declared that, if it were navigable, the North Westers peculiar to those waters and the South West monsoons of the Indian Ocean would swallow all steamers up'.

*   *   *

In fact, the Indian Navy, as the Bombay Marine had now become, was well ahead of the British in this field of steam development. At the instigation of Governor Malcolm, an astute, robust and practical Scotsman, with a lifetime of service behind him, it carried out a thorough marine survey of the channels and coast-lines of the Red Sea, with especial reference to suitable sites for coaling-stations. India itself, thanks to Malcolm, had grasped the initiative while the mother country still held back.

Sir John had been warned by the Company, on taking up his appointment, to refrain from indulging in speculative projects.

Such expensive adventures as steam navigation must be left to private enterprise. Sir John took the contrary view that only state resources could meet its high costs, and evaded his instructions with plans to develop a Bombay Government steam line. Hence the *Hugh Lindsay*, whose maiden voyage to Suez took 33 days, including 12 for coaling, and whose mails reached England in 55 days. Though this time represented good going, the ship was 'burning rupees' and the Company took the short view that the voyage was not economically justified. With an indifference inspired, so it was generally assumed, by insolvency, they failed not only to send a steamer to meet the mails at Alexandria but also to advertise the *Hugh Lindsay*'s return, which was thus even less economic. For she carried few mails and few passengers. The Company indeed sent despatches to the Indian governments, forbidding any further steam voyages to the Red Sea. But as it sent them round the Cape they arrived too late to have effect.

Over the next four years the *Hugh Lindsay* made five further voyages, which the Company virtually ignored. Her passengers, making their own way from Suez to Alexandria, had to rely on some chance sailing vessel to carry them onwards. The ship might well return empty from Suez to Bombay. For such inaction there was no valid excuse. By now the British Admiralty had established a regular steam packet service from Falmouth to the Mediterranean, and it would have been simple enough to divert it at intervals to Alexandria to meet passengers and mails from Suez. The indefatigable Waghorn, remembered as 'a very old friend of ours, and rather a troublesome one', pressed the Post Office to arrange for this. But nothing was done; it was too easy, over this question of Mediterranean transport, to retreat within the bureaucratic web of four separate responsible Government departments and the quasi-official Court of the Company. Nevertheless, at this persistent neglect of their interests, the indignation of the British citizens in India was mounting, and there was a growing insistence, through the press and the steam organizations, for some form of action from home.

Three more years were to pass before positive action materialized. The home Government was slow to move; but in its cautious and cumbersome fashion it started to do so. Mercantile organizations in Britain itself pressed, as Waghorn had done, for a change in the postage rates, which would expedite commercial correspondence via Suez, even if goods must still sail round the Cape. The House of Commons appointed, in 1832, a Select Committee to enquire into the state of trade between Britain and the East and the affairs of the East India Company. It recommended nothing, but at least Parliamentary notice was taken. British Indian 'steam agents' began to lobby intensively; petitions from Bombay and Calcutta and Madras showered down on Whitehall; and finally, in 1834, another Select Committee of the House— before which Waghorn was a witness—deemed it 'expedient that measures should be immediately taken for the regular establishment of steam communication from India by the Red Sea'.

It was decided, it is true, to give financial priority to a survey of the Euphrates route. None the less, a fundamental principle was now established. For the Government took over effective financial control of the East India Company. This committed the Company henceforward to a progressive policy, which involved the development of steam in Eastern waters. For a start it was to provide two large steamships to supplement the *Hugh Lindsay* on the route between India and Suez. As to the Mediterranean, the steam packet service from Falmouth was at last extended monthly to Malta and Alexandria, with special rates of postage for onward transmission to India. One link in this chain of communications remained to be forged—that of the overland route between Alexandria and Suez; and Thomas Waghorn, after a decade of frustration, was ready and eager to forge it.

*   *   *

Egypt had been governed for a generation by the strong hand and shrewd brain of Mohammed Ali, who had moved in to fill the vacuum left by the defeat of Napoleon and the departure of the British occupying forces. An Albanian soldier in the Turkish Army, he staged a military *coup d'état*, liquidated the Mamluks and founded a dynasty of hereditary Pashas, still under nominal

Turkish suzerainty but in fact with a wide degree of autonomy in his own affairs. He was bent on modernizing his backward country, adapting Western ideas and techniques to Oriental conditions in the economic and social fields, and establishing friendly but cautious relations with foreigners equipped to assist these ends. He encouraged trade, created industries, reclaimed lands and increased irrigation. To improve communications he cut, at great cost of human labour and life, the Mahmoudieh Canal, linking Alexandria with the Nile and so, without being aware of it, carrying out a part of the project recommended by Le Père. The Pasha was thus disposed to favour Waghorn's project for a regular route between Alexandria and Suez, provided his own interests were safeguarded; and he issued a firman accordingly.

A few weeks after the extension of the British Government's mail service to Alexandria, Waghorn announced publicly his plan to make use of it on behalf of the British commercial community in general. He established himself in Egypt as an agent for the rapid transport of mails, goods and passengers between Alexandria, Cairo and Suez. He promised a regular service, through which the mails would be transported by his own messengers, either to be embarked at Suez on ships of the Indian Navy or despatched in the care of a Maltese carrier by native boats down the Red Sea to Mocha, Aden or Socotra, and so onwards to India. Letters must be registered and paid for at a minimum rate of 5s. 5d. per ounce to India, through one of his agents in Britain, and franked with the official stamp and with his own cachet, 'care of Mr. Waghorn, Alexandria'—a mark which was soon to make Waghorn a household name throughout the East. For passengers he undertook to secure berths at Falmouth, to ship their baggage, and to arrange for their conveyance through Egypt.

Thus Waghorn's Overland Route was inaugurated. Two years later, in 1837, 'this person' achieved official recognition at last. He was appointed Deputy Agent of the East India Company in Egypt, hence in effect of his former detractors the Post Office, for the superintendence of mails; and soon afterwards all native postal agents of the Bombay Government were replaced by his own salaried employees. To reduce costs over the route, he transported British coal overland, by barge and camel, from Alexandria to Suez, where it could thus be sold for £3 instead of £10 a ton and so shipped to the Red Sea stations. Since it had previously been transported at far greater cost from Bombay, this was an economic factor which led to a rapid increase in steamships East of Suez.

Over the desert transit route for passengers and baggage, between Cairo and Suez, Waghorn found competitors in the agents of the Bombay Steam Committee, Messrs. Hill and Raven, who built two hotels, one in Cairo and the other in Suez, and a chain of rest-houses across the desert. For some years they operated in rivalry, Hill excluding Waghorn's passengers from the rest-houses and even denying them water, Waghorn in retaliation cornering the carriage horses, camels and donkeys which provided the sole means of transport. But eventually they went into partnership and the firm became J. R. Hill & Co.

In these early days, the overland journey across Egypt could hardly be advertised as comfortable. The first stage of it, from Alexandria, was a stretch of 45 miles along the Mahmoudieh Canal, between high banks burying the bones of 20,000 forced labourers who had perished while digging it, largely with their hands. Passengers travelled in horse-drawn boats, with horn-blowers in the bows to warn the local traffic. For the second stage, up the Nile itself to Boulac, near Cairo, there were sailing-boats, and later the *Jack o' Lantern*, a miniature steamer ('three donkey-power') which carried ten passengers. The movement of a single person would cause her to list heavily, and she would often run on to a sand-bank.

The cabins on all the boats were infested with vermin. 'On the lamp being lighted', wrote a redoubtable lady traveller, Mrs. Colonel Elwood, 'to my consternation, we discovered thousands and ten thousands of cock-roaches, running merrily about in every direction, and absolutely over our couches; and we also had the pleasure of finding our boat was infested by rats, which paid us repeated visits during the night.' During the day, taking a siesta, she observed that 'the ceiling was positively encrusted with flies and the floor was swarming with fleas'. The recollections of a male traveller, Mr. W. H. Bartlett, were not only of Egypt's plagues of 'bugs, fleas, cockroaches and other creatures more minute and "familiar to man"' but of human vermin, 'a crowded company of the lower class of Egyptians, and of the horrid consequences of unavoidable proximity with their filthy persons and populous garments'.

The passengers remained in Cairo until news arrived, at first by runner and later by a system of semaphore signals, of the arrival of their ship at Suez. There followed the last stage of their ordeal, the journey across the desert, where camels gave place, in time, to vans, like small omnibuses, drawn by relays of horses—but where the flies, as travellers complained, still abounded. At the rest houses, some literally lousy despite ablutions of chloride of lime, the flies settled in swarms on the breakfast dishes, making the food hard to identify; and when travellers ate regardless, they 'were continually crawling round the corners of our mouths and when we opened them for a mouthful of food, they also claimed admittance'.

The food—mutton chops, stewed fowls, roast pigeons—was tasteless and tough, while the eggs were boiled to the consistency of bullets. It was advisable for passengers to bring their own water in bottles, but there was port at five shillings a bottle, and lukewarm ale, porter and stout at a shilling and tenpence. Suez itself—'not a garden, not a tree, not a trace of verdure, not a drop of fresh water'—was a place of indescribable wretchedness, much as Napoleon had found it. Neither of the rival hotels was comfortable, but in Hill's, as a traveller recorded, 'the bedrooms are few, and the ultimate resort is the divan or a large cushioned seat of the dining-room, and the cold night air from the desert freely blowing on the sleepers from numerous broken panes of glass'.

None the less Waghorn could justifiably claim that, thanks to his efforts with Hill & Co., and to the protection of the Pasha, a route was established where 'even ladies, alone with their infants, could pass to and fro with ease and security, in that desert between England and India, as in Europe'. The 'wandering robbers' of Napoleon's time had been tamed into 'faithful guides'.

No act of violence, he could afterwards boast, had been committed against one of his convoys, while 'not one English individual has died of, or become infected with the plague in passing over it, or at Suez'.

Thackeray gave Waghorn his due, writing playfully in the early eighteen-forties of the hotel in Cairo, where 'a hundred Christian people, or more, come from England and from India every fortnight'. Describing its court full of bustling dragomans, ayahs, and children from India, 'and poor old venerable he-nurses, tending little white-faced babies that have seen the light of day at Dumdum or Futtyghur', he continues:

> The bells are ringing prodigiously and Lieutenant Waghorn is bouncing in and out of the courtyard full of business. He only left Bombay yesterday morning, was seen in the Red Sea on Tuesday, is engaged to dinner this afternoon in the Regent's Park, and (as it is about two minutes since I saw him in the courtyard) I make no doubt he is by this time at Alexandria or Malta—say, perhaps at both. *Il en est capable.* If any man can be at two places at once (which I don't believe or deny), Waghorn is he.

What, he asks, are Napoleon's wonders compared to those of Waghorn?

> Nap massacred the Mamelukes at the Pyramids; Wag has conquered the Pyramids themselves—dragged the unwieldy structures a month nearer England than they were, and brought the country along with them. . . . All the heads that Napoleon ever caused to be struck off . . would not elevate him a monument as big. Be ours the trophies of peace! O my country! O Waghorn! *Hae tibi erunt artes.* When I go to the Pyramids I will sacrifice in your name, and pour out libations of bitter ale and Harvey Sauce in your honour.[1]

Thus did Waghorn achieve a moment of immortality in English letters. He achieved it too by the momentary passage of his name into the English language. In a contemporary London pantomime

[1] *Notes of a Journey from Cornhill to Grand Cairo.*

the audience responded readily to the allusion when the entry of Prince Charming was heralded from the stage as being 'Waghornly timely'.

But Waghorn's day, by now, was nearing its end. The Overland Route was to become absorbed into the sphere of 'big business'. A new shipping company, the Peninsular and Oriental, was founded by two enterprising north countrymen, Wilcox and Anderson. Starting a mail run to the Iberian Peninsula, under Government contract, they extended it from Gibraltar to Alexandria. In 1840 the P. & O. was granted, under royal charter, the contract for the whole of the sea route to India, covering not only mails but passenger and fast mercantile traffic, and obliging them to introduce two powerful steamers on the route between Suez and Bombay.

Inevitably this led them towards the control of the Overland Route. They introduced their own boats on the canal and on the Nile, and Anderson negotiated with Mohammed Ali the formation of an Egyptian Transit Company, to operate and improve the desert route to Suez. They bought out Hill and froze out Waghorn, whose stubborn spirit of independence precluded him from coming to terms with them. For a while he tried to compete, reaching a separate and limited arrangement with the new Company. But finally he gave up the struggle, and soon afterwards, in 1846, the control of the transit route was relinquished by the P. & O. to the Egyptian Government.

Waghorn transferred his activities to Europe, trying to establish an alternative route to Alexandria which would pass through Trieste and thus by-pass France, with whom political relations looked precarious. Once again he obtained no official support and fell seriously into debt. His health broke down and he became a man with a grievance, as Baldwin—and to a lesser extent Bruce—had been. Claiming that he had sacrificed his career, his constitution, his private fortune in the public interest, he bombarded the British Government for redress and assistance. But an ungrateful nation, as he saw it to be, did little. Eventually he was granted a small pension by the East India Company and another by the Government, together with a lump sum of £1,500, which his creditors impounded. He died shortly afterwards, in his forty-ninth year, his wife surviving him for seven years in a state which bordered on destitution.

Thomas Waghorn, too impetuous in youth to command official confidence, possessed and already burnt-out in middle-age by a demon of energy, did not shake the world. But, by the unflagging pursuit of a single realizable ideal, he did help to give it new direction through one of its radial points. Credit for the opening of the steam route through Suez must go first to Sir John Malcolm, as Governor of Bombay, and to the British citizens of India themselves. But this lone, impatient, ubiquitous Englishman, doing what he had set out to do in defiance of obstacles, became part of its legend.

Another individual, who was to change the world at this point more radically, and to survive, through his creation of the Suez Canal, as a mightier legend, always admired this 'man with a craze'. The Frenchman, Ferdinand de Lesseps, saw him as a founder, through his overland route, of navigation between Britain and India and accorded him immortality in bronze by erecting a monumental bust 'To Waghorn', overlooking the mouth of the Canal at Port Tewfik. Unveiling it, he said: 'He opened the route. We followed.'

Ironically, Waghorn had strongly opposed the idea of the Canal when it arose in his day. But Lesseps, who had met the young Lieutenant, remarked of him afterwards: 'It was the courage he exhibited that left a deep impression on my mind, and served as an example'. An example to become the 'Vasco da Gama' of Suez?

CHAPTER III

Schemes for a Canal

Throughout these first decades of the nineteenth century, there was little or no serious thought of a canal across the Isthmus of Suez. Le Père, Napoleon's engineer, seemed to have said the last word on the project. In 1830, it is true, new evidence emerged which might have led to a positive change in its prospects—but it did not. An artillery officer, Captain F. R. Chesney, was sent by the British Ambassador in Constantinople to Egypt and Mesopotamia, to survey alternative overland routes through the Ottoman dominions to India, to report on all and to recommend the best. Chesney was an Ulsterman of hardy, God-fearing stamp, respected for his authority, his endurance, and his coolness in face of all danger. Reading a paper to an audience on one of his projected journeys to the Euphrates, he was asked: 'But won't there be difficulties with the Arabs?'

'Difficulties, Sir?' he replied indignantly. 'Do you think *I* would have had anything to do with it if there had been no difficulties?'

In his relations with Arabs he was irascible and stern, 'thrashing them into their work', but once pulling himself together with the reflection, 'I must endeavour to take things more quietly, and, if I cannot rule my spirit, learn at least to be indifferent to the stupidity and idleness of our servants.'

In Egypt, after climbing Mount Sinai twice in one day, he explored the traces of the ancient canal, made a general—but sketchy and not very professional—survey of the Isthmus, and sailed across Lake Menzaleh to examine its outlets to the Mediterranean and the Damietta mouth of the Nile. The fishermen of the lake made a sort of honourable prisoner of him on one of their islands. He feared serious mischief, but contrived to remain 'composed in this trying scene, smothered with dusty, noisy Arabs'—who did no more than plunder him of all but his ink-bottle.

As a result of these investigations he concluded—and reported —that 'a sea-canal could be opened, so as to give a passage for steamers and other vessels, without even so much disadvantage as is experienced in the case of the Bosporus'. In reaching this conclusion, he threw doubt on Le Père's estimate of the differing levels of the two seas:

> For whatever natural causes can be supposed to exist likely to maintain the Red Sea at a higher level, can hardly fail to influence equally the Mediterranean at the distance of little more than 60 miles. The land, it is true, shelves gradually from the Red Sea to the West shore of the Isthmus, at a mean difference of 18 ft., according to French engineers. But it is very questionable whether the sea itself is really higher, communicating, as it does already, with the Mediterranean, round Africa; but even if it could prove so an additional inlet will no more increase the height of the latter sea, than do the infinitely *more* voluminous ones passing in from the Atlantic on that side and the Black Sea on the other.

Allowing for evaporation, he forecast only a moderate current from the Red Sea to the Mediterranean—but a current sufficient to keep the canal constantly clear from shifting sands. Moreover, across the Isthmus, 'not a single mountain intervenes, scarcely what deserves to be called a hillock'. There would thus be no difficulty in cutting a canal as far as Lake Menzaleh. Finally, with regard to the cost of construction, there was Arab labour without limit.

However, Chesney admitted, 'the possibility of constructing a ship canal to Suez did not long continue to occupy my thoughts'. His thoughts—and his heart—were set on the Euphrates route. So were those of his Ambassador and of the British Government, animated to a great extent by a political motive—that of containing, through British control of Syria, Mesopotamia, and the road to Persia, the threat of southward expansion by Russia.

The Select Committee on Steam Navigation to India in 1834 adopted this policy, giving second priority to the Egyptian overland route, and brushed the canal project aside altogether. The Committee responded readily to the influence of Chesney, who argued that the opening of a Suez Canal would be 'a magnificent thing for the world at large but not so much for the benefit of this country in proportion'. For a great deal of the commerce would go no further than the Mediterranean, and thus its main beneficiaries would be the countries of Europe. Egypt, he maintained, offered Britain few political advantages by comparison with Mesopotamia. Moreover, the Euphrates route offered also the sublime opportunity of introducing Christianity into the Arabian peninsula, for 'civilization always follows in the train of commerce'. His attempt, however, to open up the route from Aleppo to Basra, by river steamer down the Euphrates and Tigris, ended in failure, owing to the hostility of Arab tribesmen. It was then that the British Government turned more serious attention to the Suez route—but still not in terms of a Suez Canal.

In 1841 Arthur Anderson, a director and founder of the P. & O., was in Egypt in connection with his company's plans for the Overland Route. He became interested in the possibility of reopening the ancient canal, largely through a French engineer, Linant de Bellefonds. Linant Bey had been in Egypt for some twenty-five years, most of which had been spent in Egyptian Government service.

A legend in his lifetime to visiting countrymen, Linant dispensed hospitality in an oriental house, with a Syrian wife, who wore the costume of her country and sat on a divan piled with purple cushions. Around them was a bevy of young girls, whom Gerard de Nerval took to be Linant's seraglio, until he learned that they were in fact his own daughters and the marriageable daughters of friends. Flaubert, on a visit to Egypt ostensibly for the French Ministry of Agriculture and Commerce, described him as the most intelligent of all the French Beys. He was Chief Engineer to Mohammed Ali, and now presided over his Ministry of Public Works, controlling Egypt's vast and complex system of irrigation and inspiring the Pasha's plans for its widespread development.

Linant had explored the Isthmus of Suez from sea to sea in

topographical and geometrical detail, making a map of it and working out a technical project for its canalization. At Mohammed Ali's request he had later studied both the direct and the indirect routes for a canal. Though he was still working on the basis of Le Père's assumed difference in levels, he yet preferred the direct route between Suez and Pelusium, moreover for a canal without sluices. For he considered that the gentle slope, which this difference implied, would make for a channel with a moderate current, deepening and clearing it and assisting its maintenance. He drew up a memorandum for the Pasha which summarized these conclusions.

The Frenchman's arguments caught the imagination of the seafaring highlander, Anderson. He pressed them on Lord Palmerston, who was then Foreign Secretary, pointing out that the physical impediments to a Suez Canal were now proved not to exist, for the ground between the two seas was almost level, moreover easy to work. The time had surely come when the political obstacles too could be overcome, and 'an object of almost universal utility, whether viewed in commercial, or political light be accomplished as a natural consequence of Your Lordship's policy'. What Anderson proposed was a 'Great Oriental Canal', internationally protected, which would benefit all nations but particularly Britain, with her political and commercial interests in India and the East. It could double the present tonnage of her Eastern trade; it should be financed 'chiefly if not exclusively by means of British capital, and would be directed by British subjects'; and it should surely yield an ample return to investors.

Anderson followed up his letter to Palmerston with a full memorandum, to be amplified in a final version under the title of 'The Practicability and Utility of opening a Communication between the Red Sea and the Mediterranean'. After asserting the axiom that 'the facility of intercourse, and commerce carries with it civilization', he summarized Linant's own memorandum with proper attention to technical detail. He increased the Frenchman's modest estimate of the cost from £250,000 to £1 million, but stressed the ultimate financial returns through the inevitable development of steam navigation.

Politically he emphasized that the consent and co-operation of Mohammed Ali would be needed if foreign capital were involved. No man, he explained, had suffered so much from European projectors and speculators—'a host of schemers and adventurers who led him into all kinds of experiments', ending often in disappointment and loss. The Pasha's rights of ownership and control would have to be asserted. He might demand a right to a share of tolls, in perpetuity. Above all, he would be unlikely to agree to the project unless through the intervention of one or more foreign Governments.

Soaring into a peroration of high-minded eloquence, Anderson insisted on the benefits of such a canal, not only to the British Empire but to the human race in general:

> Five hundred millions of human beings inhabiting Hindustan and China remain to this day enslaved by debasing superstitions, and sunk in mental darkness and delusion. What a field is here opening to the Christian philanthropist! To aid in the removal of ignorance and superstition by the diffusion of useful Knowledge and an enlightened religion, to plant industry and the arts where indolence and barbarism have hitherto prevailed, are noble efforts, tending no less to elevate those who engage in them, than the objects of their exertions. The opening of the proposed communication would obviously subserve the promotion of such objects, and therefore can scarcely fail to excite an interest in the mind of every well-wisher to his fellow creatures.

But to his far-sighted and sensible proposal Anderson received only a formal acknowledgment from Palmerston. The British Government was not interested. Arthur Anderson indeed, who held a leading position in shipping and trade, was alone among men of standing in Britain thus to grasp the full significance of a canal to unite the two seas and officially to press for its execution. At no time either now or in the predictable future did the imagination, the commercial resource, or the scientific enterprise of the British nation seriously respond to the idea.

38

The predominant reasons for this were political. The British Government's field of vision was bounded politically by fear of France and her suspected competitive designs in the East; and technically by a preoccupation with railways, at which British engineers excelled. In a safe, matter-of-fact spirit, the Government was to pursue the limited objective of a land line for steam across Egypt rather than a ship canal between the two seas. Thus inevitably the ship canal, which might have been a British enterprise, was destined to go by default to the French.

The concept of a Suez Canal was now re-born in France, and matured in the minds of Frenchmen. Visionaries, philosophers, metaphysicians conceived it once more as they had done under the *ancien régime*. Its 'father' was Henri, Comte de Saint-Simon, a wealthy progenitor of Christian Socialism, who sought the social and spiritual regeneration of mankind through the dignity and responsibility of labour. A descendant of Charlemagne, his forbear the Duc de Saint-Simon had shone at the Court of Louis XIV. He himself was a creature of the French Revolution and of the ideas which preceded it. His valet had orders to wake him each morning with the words, 'Remember, Monsieur le Comte, that you have great things to do.' One of these was the junction of the Atlantic to the Pacific; another that of the Red Sea to the Mediterranean. Both aspirations were echoed by Goethe.

Saint-Simon died in 1825, bequeathing his mission to a disciple, Prosper Enfantin. The son of a bankrupt banker turned teacher, Enfantin received a Polytechnic education in political economy and finance. He himself, after travelling for a while as a wine merchant, became a cashier in a bank. But, always susceptible to the emotional influence of cosmic ideas, he began to see himself as a Prophet, divinely inspired to create a new world to which peace and enlightenment would come, not only through noble toil but through the wonders of science and the development of industry, trade and technology.

He became the high-priest of the St. Simonian cult, gathering around him a band of Apostles who forsook all to follow him, renouncing their property to gain each day their daily bread. They made themselves conspicuous in sacerdotal tunics of blue—

39

light blue for himself as Supreme Father, darker blue for the followers, deepening to royal blue for the lower ranks of the hierarchy. In these costumes they listened to their Father's eloquent if not always comprehensible orations, finding his words more sublime than any since the Sermon on the Mount. They practised elaborate ceremonial, confessed their sins in public, and marched in solemn processions, illuminated throughout by the majesty of Enfantin's countenance and the divine brightness of his smile.

In the spring of 1833 Enfantin and thirteen of his disciples sailed on a sacred mission to the East. The hymn of their cult, which they chanted, and the dress which they wore—Enfantin in a nautical singlet with 'The Father' inscribed like the name of a ship across the chest—caused some commotion on the quay as they embarked at Marseilles, and the police had to intervene to prevent them from being thrown into the harbour.

This was a journey of deep significance for the St. Simonians. In his pursuit of universal peace, the Father aspired to unite the East, both spiritually and materially, with the West. In the terrestrial sphere this was to be achieved not in political but in social and economic terms: by the introduction of modern Polytechnic education into Egypt, and by communications—the cutting of a canal across the Isthmus of Suez. 'It is for us', declared Enfantin, 'to establish between ancient Egypt and old India one of the new routes from Europe to India and China; later we can also cut the other at Panama. We shall then have one foot on the Nile, the other on Jerusalem, our right hand extended towards Mecca. Our left arm shall cover Rome and yet rest on Paris. Suez is the centre of our work and life.'

In the celestial sphere Enfantin believed in a deity who was not only God the Father but God the Mother. There was need for her material incarnation in the form of a female Messiah with whom he, as the incarnate Father, would unite; and he hoped now to find her in the East. The East represented the female principle in humanity, the West represented the male. Thus he sought a marriage between them—a marriage made in Heaven but symbolized on Earth by the penetration of the Isthmus of

Suez. 'There', he wrote, 'we shall perform the act which the world awaits as a proof that we are males.'

In terms of flesh and blood, this principle presupposed the equality of woman with man. Hence free love reigned among the St. Simonians. Enfantin opposed the 'tyranny of marriage', sanctioning successive unions with different persons for the happiness of those disposed by nature to change and, for their benefit, supporting easier divorce. The flesh was as sacred as the spirit. Natural passions, the gratification of the senses, must be given free if temperate scope. These principles were to lay his followers open to charges of subverting morality. Meanwhile they were to be tried out in the female East.

In Constantinople the St. Simonians, in their outlandish costumes, rendered open homage to womanhood, 'speaking out loud and with hats off to the daughters of the East, poor and rich, on foot and in carriages', saluting every woman they saw in the street. Such practices shocked and alarmed the Turks, who were far from ready, at this stage in their social history, for female emancipation. Thus Enfantin was arrested and conveyed to a prison in Smyrna.

Here he prudently decided on a change of tactics. For the next stage of the long journey to Alexandria the disciples abandoned their apostolic garments and shaved off their beards. Women, for the moment, were to be forgotten. Mystical enthusiasms were to be harnessed to practical tasks. The emphasis was to be on industrial and political matters. Enfantin, nurtured on the works of the savants who accompanied Napoleon to Egypt, had himself enrolled a small group of engineers and mechanics with Polytechnic diplomas.

The man who met them on arrival was Ferdinand de Lesseps, the French Vice-Consul in Egypt, whose father, Mathieu, had for long been a respected Consul-General before him. Arriving there from Tunis in the previous year, the young Ferdinand had been delayed for many days in the quarantine station at Alexandria, owing to a suspected case of cholera aboard his ship. Among the books brought to him from the Consulate, to help pass the time, was Le Père's memorandum on the junction of the two seas. This appealed to his imagination and kindled a resolve to study the question in all its aspects. Thus he was predisposed to help the St. Simonians, and if necessary to intercede on their behalf with Mohammed Ali. On taking up his post Lesseps had been received by the Pasha, who said to him: 'It was your father who made me what I am. Remember you can always count on me.'

Enfantin hoped to secure from Mohammed Ali a concession for the construction of a canal. He made a brief tour of the Isthmus, whose desert spaces filled him with a general awareness of God, and sped him on with his sublime mission of canalization. His engineer, Fournel, expounded its objective in a more concrete sense to Mohammed Ali in person. But he found the Pasha preoccupied rather with the question of railways, and could only gain his attention by presenting him, a day or two later, with a small-scale model for a railroad between Cairo and Suez.

Otherwise the Pasha's favoured public works project was for a great barrage across the Nile below Cairo, to regulate its flow and to irrigate the lands of the Delta. This appealed to Enfantin, economically as a means of disseminating the Nile waters for the benefit of his Egyptian 'chosen people', more romantically as a means of reviving, in a new city at their point of intersection, the glories of the ancient capital of Memphis. He hoped to have a hand in the construction of the barrage, but this task was entrusted to Linant, the Pasha's own French engineer. Meanwhile the plan for a canal was shelved.

This spelt defeat for the St. Simonians, whose group of technicians began to disperse. One by one, led by Fournel himself, they returned to France, while others succumbed to an outbreak of the plague. Enfantin himself made a long trip to Upper Egypt, here finding 'isolation from a world in which my presence has become useless and often injurious'. His dreams had turned to chimeras. Even the female Messiah had failed to materialize. Alone by now, he returned to France in 1836, spiritually a sadder but materially, as it was to prove, a wiser man.

*　　　*　　　*

Enfantin had remarked of Mohammed Ali that the Suez Canal would be made as soon as he thought a little less about his armies. For the present his armies remained the Pasha's principal concern. Not content to cultivate his garden of Egypt, he turned to a career of territorial expansion. In 1831 he embarked on an invasion of Syria—as his hero Napoleon had done before him and the Pharaoh Ramses II before that. The temptation to do so was great, since the Ottoman Empire was disintegrating and he saw a chance of throwing off his vassalage to the Sultan, ruling as an independent sovereign and passing on to his descendants the country of his creation, which would otherwise fall back at his death under Ottoman misrule.

His son Ibrahim Pasha, who was placed in command of his armies, met with little resistance. He occupied Syria and penetrated beyond the Taurus mountains into Anatolia, in the heart of the Turkish homeland. His conquests as far as the Taurus were conceded in a peace Convention which the Sultan had little choice but to sign. This made Mohammed Ali master of a sizeable slice of the Ottoman Empire, still only as a Pasha and a nominal vassal of the Sultan, but now secretly nurturing dreams to reign in his place over an Arab Empire, from Constantinople. His success initiated a critical decade in Middle Eastern affairs in which the four great powers of Europe, with their respective designs on the crumbling Empire, became closely and intricately involved.

To preserve European equilibrium, they were confronted in essence by two alternative policies: that of dismembering the Ottoman Empire, and that of conserving it. Russia favoured dismemberment, seeing the Empire as a natural outlet for her own imperial expansion. Britain favoured partial dismemberment in Europe—as lately with the amputation of Greece—but conservation in Asia. She sought above all to protect her own Empire in the East against Russian expansion, and believed that Turkey could be strengthened and preserved from decline by internal reforms. Austria, like Britain, supported conservation. France, in principle, supported it too, but favoured partial dismemberment in Asia—and here now was her opportunity. Let

Mohammed Ali amputate Syria, which could then become a sphere of French influence.

Thus the French, always competing in Egypt for the Pasha's favour at Britain's expense, secretly encouraged his invasion and played an influential part in the diplomacy which led to his Convention with the Sultan. In France his campaign had stirred popular enthusiasm. The Paris of Louis Philippe was in a Napoleonic mood; the land of Egypt awakened dreams of glory, and Mohammed Ali, marching into Syria in Bonaparte's steps, became a romantic conquering hero, whom the King himself saw as a nineteenth-century Alexander the Great. And if he turned out to be an enemy of Britain, so much the better. On all sides the French heart ruled the head at the expense of real French interests, and in contradiction with the means of achieving them.

For inevitably the French, in encouraging Mohammed Ali, had also encouraged the Russians, whom they sought to restrain. As quick to seize their own opportunity, they responded with alacrity to a request for aid by the Sultan, sending Russian engineers into Turkey to construct fortifications, officers to train the Turkish armies, troops to reinforce them in Asia Minor, and finally a Fleet to the Bosporus. Soon after the peace Convention they signed a treaty of their own with the Sultan, receiving as the price of their support virtual control of the Straits, and a position in his country which could amount to a protectorate. For a while Constantinople was to all appearances a Russian city. 'It is manifest', commented Lord Ponsonby, the new British Ambassador in Constantinople, 'that the Porte stands in the relation of vassal to the Russian Government.'

The powers could not allow it to remain so. France, having gone too far, momentarily joined Britain in a protest to the Porte, and there was even talk of war. Tension was relieved when the Tsar renounced any intention to take advantage of his rights under the treaty, and withdrew his forces. Palmerston saw, for the present, no further danger: 'With Russia', he wrote, 'we are just as we were, snarling at each other, hating each other but neither wishing for war. . . . Our policy as to the Levant is to remain quiet, but remain prepared.' But according to Bulwer, a

Secretary of the British Embassy, the British and French Ambassadors in Constantinople would go to their respective windows on rising out of bed—Ponsonby at six o'clock in the evening, Roussin at six o'clock in the morning—'prepared to see without surprise the Russian Fleet anchored under their eyes'. Suspicion of Russia was 'perhaps the only point on which these representatives of the two countries agreed'.

Mohammed Ali meanwhile looked upon his Convention with the Sultan as no more than an armistice. He was still bent on the final stage of complete independence, with a hereditary dynasty, and prepared for a renewal of hostilities, with French connivance and in disregard of British warnings. It was the Sultan who was finally goaded into taking the offensive. In 1839 his armies invaded Syria, and were routed by Ibrahim; his Fleet sailed to Alexandria, and surrendered to the Egyptians.

Britain was thoroughly roused by the Turkish defeat. Here now was a double threat to her imperial interests. Her communications with India had always, as she saw it, been menaced by Russia through Persia and Afghanistan; now they were menaced by France, through her dominance over an independent Egypt and Syria under French protection.

Hitherto Palmerston, in his policy of conserving the Ottoman Empire, had tried to maintain a balance between the powers by leaning towards France to check Russia. Now, unless the French would agree to coerce Mohammed Ali instead of inciting him, he must lean towards Russia to check France. To Lord Granville, his Ambassador in Paris, he had written: 'We ought to support the Sultan heartily and vigorously; with France if France will act with us; without her if she should decline.' Since France manœuvred and temporized, he now moved without her, boldly acting with Russia and with Austria and Prussia, whose chief concern was to maintain the existing Balance of Power. In 1840 these four powers agreed on a Convention for a pacific settlement of the Eastern Question, and signed it without the knowledge of France. It confirmed Mohammed Ali as Pasha of Egypt, still under Ottoman suzerainty, on a hereditary basis, and of southern Syria—but for his lifetime alone.

In France there was a general outburst of fury at this underhand *coup* by *Perfide Albion*. The streets rang with the 'Marseillaise' and an angry mob attacked Lord Granville's carriage. Smarting at the humiliation, the French Government called up its reservists and fortified Paris. Britain resounded with rumours of war. But Palmerston discounted the threats of the French as mere 'swagger and bully', and robustly instructed his Ambassador to convey to them, 'in the most friendly and inoffensive manner possible that if France throws down the gauntlet we shall not refuse to pick it up; and that if she begins a war, she will to a certainty lose her ships, colonies, and commerce before she sees the end of it; that her army of Algiers will cease to give her anxiety, and that Mehemet Ali will be chucked into the Nile'. Lamartine, the French poet and statesman, who had recently visited Syria, was warning his countrymen in similar terms: 'England will accept a century of war in the Mediterranean with us and the whole world rather than concede the keys of Suez to a legitimate sovereign supported by the hostile influence of France in Egypt.'

Meanwhile it was Britain who went to war—against Mohammed Ali, who refused to accept the Allied Convention. An Allied squadron, under Commodore Charles Napier, bombarded the port of Beirut, and landed British Marines and Turkish troops, who united with Arab rebels in Syria to cut off Ibrahim's soldiers and put them to flight—'Like the flock of sheep', as Bulwer put it, 'which Don Quixote mistook for an army.' The fortress of Acre, which had resisted Napoleon, fell. 'Napier for ever!' exulted Palmerston to Granville. In Paris, Heine declaimed that 'the thunder of the cannons of Beirut finds its echo in every French heart. The storm approaches closer and closer.' The French, remarked Greville, reacted as though the British had bombarded Boulogne or Toulouse. But no storm broke. British naval power had disposed of Mohammed Ali's 'invincible' army in Syria as, a generation earlier, it had disposed of Napoleon's. The French, in Palmerston's words, 'are beat, and there's an end of the matter'.

The Sultan deposed Mohammed Ali. Still counting on French intervention the Pasha put up a last resistance. But Louis Philippe changed the French Ministry, replacing the belligerent Nationalist Thiers with the more pacifically-minded Guizot, and declared for peace. Mohammed Ali surrendered. The French intervened to save his face and their own in Egypt, and here the British were now disposed to co-operate. Guizot, seeking to propitiate them, suggested the neutralization, on the analogy of the Straits, of all routes to India between the Mediterranean and the Indian Ocean, whether by the Euphrates Valley or Suez. But this enlightened proposal was brushed aside.

So, on the other hand, was an initial British proposal to limit Mohammed Ali's power by various political, financial and military restrictions. In the end the British Government came round to the view that a semi-independent Egypt, no longer under French protection and still under nominal Turkish suzerainty, might after all suit British interests best. Thus in 1841, a firman from Constantinople revoked Mohammed Ali's deposition and restored him as Viceroy, paying tribute to the Sultan with powers over Egypt alone but with hereditary sovereignty. This was confirmed in the form of a Convention signed by the powers, who now included France. The Eastern Question was thus momentarily settled, to the satisfaction of the powers of Europe.

*   *   *

Such was the framework of events within which the battle for and against a Suez Canal was to be fought in the years that followed. Henceforward an unsettled climate prevailed over Anglo-French relations in Egypt. Largely because of it, the British Government now threw its full weight on the side of a railway between Alexandria, Cairo and Suez. A ship canal, if it proved to be practicable—which Palmerston airily doubted—was a global conception. It would change the geographical status of Egypt and raise major international issues. A railway, built and operated by Britain, would serve her own transit purposes quite as well, and could be classed as a purely domestic enterprise, free from such widespread political complications.

The project for a railway had first been considered in 1834, when Galloway Bey, a British engineer in the Pasha's service, went on his instructions to England to order rails and equipment and to obtain the British Government's formal consent. This was withheld, owing to the disturbed situation at the time. Now, soon after the settlement of 1841, Britain revived the plan for a railway, offering the Pasha finance for its execution. In February 1847 Palmerston instructed his Consul-General, Charles Murray, to put its advantages before him.[1] As to the alternative plan for a canal, he was to 'lose no opportunity of enforcing on the Pasha and his ministers the costliness, if not the impracticability of such a project', and to add that 'the persons who press upon the Pasha such a chimerical scheme do so evidently for the purposes of diverting him from the railway which would be perfectly practicable and comparatively cheap'. These persons were once again the St. Simonians, whose former mystical enthusiasms were now being diverted into hard-headed industrial channels.

Enfantin, encouraged on his return to France by the growing public interest in the idea of a canal, set his followers to work on statistics of European trade with the East, and the prospects of international financial backing. Now a respected citizen and a member of the administrative council of the P.L.M. railway, in close touch with the bankers of Lyon, he embarked on a voluminous correspondence, to gain the support of banking, trading and engineering circles in the various countries of Europe.

He found a ready ally in the Austrians, now awakening to the benefits of such a canal to their Mediterranean trade. Linant's report, which had prompted Anderson's neglected proposal to Palmerston, had been shown to Metternich. 'I consider the canal', he wrote, 'as a world event of the first importance . . . and place it among the occurrences which mark epochs of great development. I am convinced that it will open a future . . . I shall do everything to induce the Pasha to occupy himself with the Suez Canal, and with the elimination of everything which stands in the way of that enterprise.'

Much stood in its way. To the confidential overtures which

[1] Later Hon. Sir Charles Augustus Murray, K.C.B.; second son of the 5th Earl of Dunmore.

followed, Mohammed Ali responded with cautious interest. He made it clear that he did not desire the participation of Turkey in any scheme for a canal; and Metternich saw this as a serious if not an insuperable obstacle. On the other hand, the Pasha was open to any suggestion, such as the Austrians made, that the scheme should be backed jointly by the European powers, guaranteeing him ownership, construction costs and tolls, and accepting responsibility for protection and upkeep. This, as Anderson had anticipated in his report, was the Pasha's minimum requirement.

But, from his recent bitter experience of Anglo-French rivalry, he doubted its political practicability. To the agent of a German society, formed to study and propagate the idea of the canal, Mohammed Ali exclaimed: 'Yes, Austria and France may desire the Canal, but England, but Russia!' He concluded: 'Let the Great Powers come to an understanding and demand it of me and I am prepared to execute it. Egypt does not lack men; I can employ my whole Army.' But he foresaw and feared strong opposition from Britain.

The international approach was now, for the first time, given practical shape. Pursuing, no longer as a misty ideal but as a concrete objective, the idea of 'a kind of Bosporus in the Suez Desert', Enfantin formed a *Société d'Etudes du Canal de Suez,* with its headquarters in Paris. It was a tripartite Study Group, its component parts furnished respectively by France, Britain and Austria. Its aim was to study on the spot the possibilities of a canal across the Isthmus, to determine its technical problems, and to estimate its cost. Passing from the sphere of study to the field of execution, the canal would become the first great international and industrial enterprise of the nineteenth century, financed by European capital, independently of the European Governments, and with Egyptian participation.

Though the ultimate success of the Study Group must depend on Mohammed Ali's readiness to grant a concession, Enfantin omitted to inform him of its formation or of the imminent arrival of its members in Egypt in the spring of 1842. Prone as the Pasha now was, in his old age, to fears and suspicions of foreign plots against his independence, he viewed the enterprise coldly—the more so from the belief that it was the intention of the promoters to apply over his head for Turkish sanction. He did not, however, obstruct the engineers in their work, and even contributed towards their expenses.

The engineers leading their respective national groups were Negrelli for Austria, Paulin Talabot for France, and Robert Stephenson for Britain. In their capable professional hands, the union between East and West, so Enfantin pronounced, was 'no longer a theory, it is a business'. The work of the three groups was divided into three sectors. Stephenson and his party reconnoitred the shores of the Gulf of Suez—not an arduous task, since the Gulf had been fully charted, in recent times, by the Admiralty and the East India Company. Negrelli reconnoitred the shores of the Bay of Pelusium. The land survey of the Isthmus itself was entrusted to Talabot. He was to collaborate, largely at the instance of the Pasha, with Linant, who was after all, through his own previous researches, the Canal's true begetter.

Friction arose between Linant and the St. Simonians, as a result of which he withdrew from the Study Group. This did not however prevent him from co-operating fully with Bourdaloue, the engineer designated by Talabot for the task. Together they carried out, with a body of experts, a minute and methodical survey of the Isthmus. They subjected it to the strictest mathematical and scientific tests, with six verifications, such as Le Père's expedition, in the conditions of his time, could not achieve, and no other engineer—not even Linant himself—had seriously attempted since. Its findings, pieced together with those of Negrelli and Stephenson, were incorporated, in 1847, in a report by Talabot, which established once and for all that only a negligible difference of level existed between the two seas. Laplace and Fourier, Le Père's critics, were vindicated. All that had been unknown or half-known in the past was now truly and scientifically known. Was the dream at last to become a reality?

Not yet. In the first place Mohammed Ali refused to accept any scheme of a foreign company without foreign government sponsorship. In the second place, Talabot's conclusion provoked differing views. Talabot himself, substituting new errors for old, maintained that a canal between two level seas would be impracticable. Whereas it had been argued before that the difference in levels would make its current too swift, he argued now that the similarity of levels would make it too slow, and too hard to maintain at the required depth, not only in the channel itself but at its outlet in the Bay of Pelusium. Thus he recommended instead a new version of Le Père's old plan for an indirect canal, equipped with sluices, via the Nile from Alexandria to Suez; and the majority of the Study Group members agreed with him. Only Negrelli still favoured the direct canal, as Linant had previously done.

As for Robert Stephenson, he favoured no canal at all. Writing off the whole project in his mind, in response to Talabot's verdict, he paid up from his own pocket his share of the Study Group's expenses, and became henceforward an implacable public enemy of the scheme. His attitude was not wholly surprising in the son of George Stephenson, the inventor of railways. He had worked with his father on the 'Rocket', had followed in his footsteps as an engineer for railway lines and bridges, and became financially involved in various railway interests. Four years later, after two further visits to Egypt, he was to land a contract for the construction of the line from Alexandria to Cairo. In this context at least, his trip with the Canal Study Group had proved to be a useful reconnaissance sortie. While working with it, he received an unbalanced letter from Waghorn, describing a ship canal through the Suez desert as 'one of the most uncalled for schemes I ever yet have heard broached'. He dismissed it as a pitiful emanation from the brain of Mohammed Ali, 'an old man who being a semi-barbarian fancies that he is greater than the Pharaohs of old or Alexander the Great and the like'.

Stephenson's views coincided with those of his Government. Murray, the British Consul-General, reported to London on the Study Group's plan, and sought renewed instructions. He gave his own opinion that the Canal, if it proved to be feasible, 'would doubtless more materially affect the commercial and political

E

interests of Great Britain in Egypt than any other change which scientific enterprise could effect in the physical structure of the earth'. Its influence over the interests of India would be vast, and in the present advanced state of science he dared not say that it would prove to be impracticable.

The Foreign Office instructed him to take a passive line and to say that he had no instructions, one way or the other. Her Majesty's Government still held that the advantages of a canal could be attained nearly as well by a railway, and cheaper. The cost of a canal would be great; there would be need for an artificial harbour, which would tend to fill up with sand; so would the canal itself. It would, on the other hand, be 'a bold thing to affirm that these difficulties may not be overcome by modern engineers'. Thus Murray should not oppose the scheme absolutely.

Falling in with this guidance, he communicated some weeks later an unfavourable report—which was to prove erroneous—on the results of Austrian soundings off the coast of Pelusium. From this he concluded that 'we can safely number the Suez Canal among the most visionary projects of the day'. This accorded with the views of Palmerston, who wrote to Lord Cowley, his representative in Constantinople: 'I believe the scheme to be impracticable except at an expense which would be wholly out of proportion to any advantage to be derived from it.'

Thus plans for the railway went ahead. Meanwhile, Mohammed Ali's mind began to fail. After the death of his son Ibrahim, Abbas Pasha, his grandson, and, under the terms of the new imperial edict his successor-to-be, took over as Regent. In the spring of 1849 he received a British delegation, led by a former Lord Mayor of London, and including P. & O. directors, who put before him, as though sure of its acceptance, a scheme to finance the construction of a railway line from Alexandria to Cairo. Abbas refused to grant any such concession to foreign capital, and dismissed the delegation 'without a friendly smile'.

A few months later Mohammed Ali died, at the age of eighty. If, in his attempt to arouse Egypt 'from the sleep of ages', he had but started her on the road towards westernization, he had for better or for worse introduced her into the orbit of the Western world, with its fiercely contending economic and political interests. In the next phase of this contest, Abbas was to support the interests of Britain in preference to those of France, authorizing against French opposition the first stage in the construction of the railway between Alexandria and Suez. But he was opposed to a canal, and throughout his reign the project lay relatively dormant.

In any event, from the start of the momentous year 1848, Europe herself became distracted by her own revolutionary turmoils. It was only with the dispersal of these clouds, followed in 1854 by the death of Abbas and the succession as Viceroy of his uncle, Mohammed Said, that a new prospect dawned upon the Isthmus of Suez. Its harbinger was Ferdinand de Lesseps.

CHAPTER IV

A Concession for Lesseps

Count Ferdinand de Lesseps came of a family known in the Basque country of France since the end of the fourteenth century.[1] For two centuries past they had been settled in Bayonne, a prolific breed of provincial notables, often with an adventurous streak, producing sea-captains and even corsairs, and in the eighteenth century rising in the world to serve their country with eminence in official positions overseas. Ferdinand's grandfather, Martin, sired a line of professional diplomats. His father, Mathieu, spent his life in consular posts abroad, mostly on the Mediterranean seaboards of Europe and Africa.

Ferdinand himself, born in 1805, spent his childhood in Italy and his boyhood back in France, where on completing his education he chose to follow his father's career. After two years of diplomatic apprenticeship with an uncle in Lisbon, he accompanied his father to Tunis where, at the time of the French occupation of Algeria, Mathieu de Lesseps was serving as Chargé d'Affaires and where, still at his post, he ended his days. In 1832 Ferdinand was appointed first as Vice-Consul, then as Consul, to Egypt, where his father had served through the years of the rise to power of Mohammed Ali, after the evacuation of Napoleon's army.

Ferdinand had grown into a robust young man of twenty-six, with a gay disposition, an exuberant charm and an infectious zest for life. Already in love with Africa, he became enchanted with Egypt, cantering across its deserts on horseback, shining in the social life of the small but intimate French colony, and above all responding to Turk and Egyptian, learning their language, and

[1] They were possibly of Scottish origin, settled in Guyenne since its occupation by the English.

with his quick curiosity and lively perceptions coming to understand their ways of life and their habits of mind.

A notable Egyptian whom he befriended was a youth in his
early 'teens, Prince Mohammed Said, the fourth and, it was said,
the favourite son of Mohammed Ali. His father, bent at this time
on developing his Navy, was having him trained as a seaman.
The boy, however, was immensely fat, and though he was made
to climb masts for two hours a day, to row, to jump, to trot
round the walls of Alexandria, and to perform other rigorous
exercises, he grew no slimmer. His father, loath to send him to
Europe to finish his training, as he wished, in his present un-
sightly condition, put him on a drastic diet, strictly limiting his
meals and decreeing that he was to be accurately weighed each
week. The latest weight was transmitted to his father in Cairo,
with his weekly educational report, and, as a foreign visitor
remarked, 'an inevitable overflow of spleen invariably ensues if
the pounds are not *decrescendo*!'

Mohammed Ali permitted Said to visit Lesseps, the son of his
old friend Mathieu, and this he did frequently, making himself
at home in the Consulate's domestic quarters. He would come in
tired and famished, flinging himself on the divan to rest, and
conspiring with the servants to bring him plates of macaroni
to relieve his hunger—and so put back any pounds he had lost.
The young Prince became deeply attached to Lesseps, who took
him for rides in the desert, taught him fencing, and imbued him
with French ideas and culture. He thus laid the early foundations
of a friendship which was to help him make history.

After five years in Egypt and a short spell in Holland, Ferdinand
de Lesseps was posted to Spain. Here he had his first direct
experience of those internal convulsions which were rending the
countries of Europe at this period of change. Here was the type
of situation which Lesseps, as a man of progressive outlook
reared in a Catholic tradition, was well equipped to understand.
He served first in Malaga, seeing a revolution in which a military
junta overthrew and exiled Queen Christina, and finally in
Barcelona, where he held a responsible post as Consul. Here he
was to see revolution on a more turbulent scale, when the

separatist Catalans, thirsting for independence, rebelled against the despotic central government.

The sympathies of Lesseps, himself a creature of the revolutionary age, with a passionate belief in the freedom of the individual and a hatred of absolutism, were with the Catalans against a reactionary régime which he saw as 'entirely oriental'. But, when an insurrection broke out, it was his diplomatic task to steer a peaceful course between these two extremes. This he did successfully, serving as an intermediary between the rebels and the Government forces, and thus staving off the city's bombardment until a third of its population had been safely evacuated. For his humanity, his courage and his resource he became a hero in France—and also in a large part of Spain, where French prestige had soared. Back in Paris he had his first taste of fame, in the form of a glowing public tribute from Guizot. He then returned to Spain for a spell as Minister in Madrid.

But a major reversal of fate lay in store for him. In 1849, before taking up a new post in Berne, he chanced to attend a stormy session of the National Assembly in Paris. In the wave of agitation for national freedom which had broken over Europe, the Italians, inspired by Garibaldi, had rebelled against Austria, capturing Rome, establishing an anti-clerical Roman Republic, and driving the Pope into Neapolitan territory.

The French Republic, of which Louis Napoleon, the nephew of Bonaparte, was now Prince-President, sought on the one hand to conciliate its own Catholic people, whose power it feared, and on the other to forestall the ambitions of Austria. Thus its Assembly sent an army to Italy—the first major military intervention in Europe by France since the Battle of Waterloo. As an intervention by one democratic power against another to restore a reactionary theocracy, Napoleon's enterprise was fraught with contradiction. Its ostensible aim was the liberation and protection of Rome, not aggression against it. But inevitably bloodshed occurred, creating a furore from the Left in the National Assembly.

From a confused debate there emerged only a desire to resolve the conflict by peaceful means. There must be negotiation with the Roman Republican Government. Ferdinand de Lesseps had listened to the debate from the gallery, familiar with such issues in Spain, and deeply concerned with them as they now confronted Italy, a country where he had lived in his boyhood. Next morning he was summoned by the Minister of Foreign Affairs, Drouyn de Lhuys, and invited to proceed, as Minister Plenipotentiary, to Rome. Who better to tackle the ungrateful task of reconciling the Papal and Republican interests of Rome than this diplomat of ability and recent renown, respected alike for his liberal principles and for his Catholic beliefs? Lesseps, passionately engaged, eager for action, contemptuous of the probable hazards, accepted the mission without hesitation. His instructions were highly ambiguous, and Prince Louis Napoleon, who received him before his departure, failed to make them clearer, saying only that he must uphold French interests. His Foreign Minister said confidently: 'Your clear and straightforward judgment will inspire you according to circumstances.'

Ferdinand de Lesseps, caught up in the inescapable pattern of this Nationalist age, faced once more a conflict between Left and Right, such as he had helped indirectly to contain in Barcelona. But in Rome the issues were more complex and of European-wide import. Moreover they called for his own direct responsibility. He faced the situation with resolution and good temper, maintaining through every reverse that optimism which was an inherent part of his nature. He worked patiently for an agreement with Mazzini, the Nationalist leader, which would maintain the spiritual but modify the temporal power of the Pope. He eventually achieved it—but was forestalled by his own countryman, the impatient General Oudinot, whose forces, despite the pleas of Lesseps, marched into the city.

Lesseps returned at once to Paris, still hoping to get his treaty agreed by the French National Assembly. But in his absence elections had brought back a Right Wing majority, and a new Government bent on a policy of aggression in Rome. When a storm broke once more from the Left, at this gross betrayal of faith, the Right unscrupulously side-stepped it by turning on Lesseps and making him scapegoat. His treaty was denounced as a disgrace to French honour, and he was arraigned before a Council of State, of dubious legality.

No charges more specific than disobedience to instructions could be preferred against him, and these he rebuffed with conviction and clarity. Insinuations of disloyalty however abounded, and it was even alleged in the Press that he was a sick man, deranged by the climate of Italy. The Court could only reprimand him. His conscience was clear. But his career, at the age of forty-four, was broken. He resigned from the public service and retired into private life.

Ferdinand de Lesseps had fallen a victim to party politics—in a sense to his own political innocence. Inspired by enthusiasm, idealism, patriotism, and a belief in his powers of diplomatic persuasion for a cause he believed in, he had plunged almost on impulse and without thought of self-preservation into the shoals and cross-currents of an equivocal political venture. A more cautious man, a man more conversant with the shifts of home politics at this time of confusion, would have been chary of accepting without reservation an assignment so vague and so liable, through its inherent inconsistencies, to involve him in failure. Lesseps had accepted it. He had succeeded in diplomatic, but had failed in political terms. An excess of self-confidence and of his native optimism had undone him professionally.

But his faith in life and his sense of its true values survived unshaken. Through the next four years he lived in contentment the life of a French country gentleman, with his wife and his children, at La Chénaie, an old manor house in the Berry country which his wife's mother had bought, and which he now restored. He studied agriculture and aspired, by up-to-date methods, to create a model farm on its lands. Hard hit by the death of his wife in 1853, as he neared fifty, he began to hanker once more for action, and for the world he had left, with its manifold problems.

Throughout these years of retirement and leisure, he had remained fascinated by the idea of a canal across the Isthmus of Suez, re-reading and pondering again on the reports of Le Père and of Linant, which had kindled his youthful excitement in Egypt. In 1852 he drafted a memorandum on the subject, which he sent to Ruyssenaers, the Dutch Consul-General in Alexandria. 'I have never ceased', he wrote, 'to study in all its aspects a question which occupied my mind when we became friends in Egypt twenty years ago. I admit that my venture is still in the clouds, and I don't deny that as I alone think it possible, it may prove to be impossible.' Ruyssenaers replied that there was nothing to be done for the present, owing to the hostility of Abbas to any projects from Europe and from France in particular.

Lesseps then approached the financier Benoît Fould, who planned to found a banking-house in Constantinople. Seeing the advantages of the canal scheme, Fould sounded the Turks, but received the answer that this was a matter for the initiative of the Viceroy of Egypt alone.

Then, one day in the autumn of 1854, Lesseps was busy at La Chénaie with masons and carpenters, building an extra storey on his manor house, when the postman arrived with the mail from Paris. Breaking off, he read the momentous news that Abbas had died and had been succeeded as Viceroy by—as Lesseps put it in another letter to Ruyssenaers—'that friend of our youth, the intelligent and warm-hearted Mohammed Said'. He at once wrote to Said, complimenting him on his inheritance and saying that, as he had himself now had leisure from politics, he would willingly visit him in Egypt to pay his respects in person. Said replied, giving him a rendezvous in Alexandria for the early part of November. 'What good fortune', Lesseps concluded to Ruyssenaers, 'to find ourselves together again on our old ground in Egypt! Before I arrive, not a word to anyone about the project of cutting the Isthmus.' Suddenly the clouds had cleared from the horizon of Ferdinand de Lesseps. His way ahead was clear.

Already he had been in touch with Enfantin and the members of his Study Group. Two at least of them—Arlès, a business man from Lyon, and Dufour-Férance, a Leipzig banker—had pressed 'the Father' to encourage his participation. But the tactical approach of the St. Simonians was radically different from that of Lesseps. They sought first to obtain money from Europe, then to obtain the blessing of the European Governments, and then only to seek a Concession from the Governments of Turkey and Egypt.

Lesseps worked the other way round. The key to the project as he saw it was Egypt. First, he insisted, secure a Concession from Egypt; then bring in the Governments and the bankers of Europe.

And now at last he had the chance to pursue in practice the course which intuition and reason had told him was right. On the advice of Arlès he stopped at Lyon on his way to Alexandria. Enfantin, despite reservations as to Lesseps's approach to the problem, handed over to him relevant documents of his Study Group, including the maps and memoranda of Linant and Talabot.

*   *   *

It was with emotion that Ferdinand de Lesseps landed once again on the shores of Egypt, the country he had come to love in his youth. He was met by his old friend Ruyssenaers and by an emissary from the Palace, who placed at his disposal, on behalf of his master, a spacious and luxurious villa on the banks of the Mahmoudieh Canal. Here he was saluted by a large staff of servants who included, for his especial comfort, a Greek valet and a Marseillais cook. In its garden he breathed again the jasmine fragrance of Egypt, refreshed by a light breeze from the waters of Lake Mareotis. Said Pasha—who had not perceptibly thinned with age—received him warmly, recalling memories of his boyhood, when Lesseps had helped to mitigate his father's severities, and the persecutions he had since suffered at the hands of his nephew Abbas. Now that Providence had confided to him a power which Lesseps chose to define to him as 'the most absolute on earth', he was resolved to do good for his country and enhance its prosperity.

Lesseps saw him each morning, whether officially or otherwise, but chose not to precipitate the question of the Canal. Mohammed Said once said, in the hearing of Ruyssenaers, that his father had renounced the piercing of the Isthmus for fear of trouble from Britain, and that he himself, on his accession, would follow the paternal example. Still Lesseps, characteristically, remained sure of success.

The Viceroy presented him with a fine Arab horse, brought from Syria. Mounting it, he rode across the desert to join him at a review of his troops. While they galloped together, a tassel encrusted with diamonds fell from the Viceregal cartridge-box, but Said rode on, permitting no one to stop and retrieve it. Next day he gave orders for his troops to set out for Cairo, and invited Lesseps to join him at his desert headquarters. In camp they slept on iron bedsteads beneath silken quilts, washed their hands in silver basins, and after a sprinkling of rose-water sat down to a seven-course meal, eaten with the fingers by all except Lesseps, for whom a *couvert* was laid, with porcelain-handled cutlery and plates of Sèvres. After dinner he sowed in the mind of the Viceroy the idea of starting his reign with some grand and useful enterprise.

Together they rode across the desert, at the head of the Viceregal troops, who looked healthy on a ration of three biscuits a day. Lesseps chased a jackal. The Viceroy watched his artillery at target practice. A gazelle hunt was promised. At night, beneath a full moon, the Viceregal band played the tunes and marches of all countries, not forgetting the 'Marseillaise'. Around the camp fires the Egyptians, 'the gayest people on earth', just as Lesseps remembered them, sang their Arab songs, standing in circles and rhythmically clapping their hands.

Next morning Lesseps awoke early. Standing before his tent he looked to the right and saw the dawn breaking slowly over the eastern horizon; he looked to the left and saw still only darkness and cloud. Then suddenly out of the gloom there leapt up from the western horizon a rainbow which swiftly bestrode the sky, in all its shining colours. The heart of Lesseps beat strongly. His imagination caught fire. This arc across the Heavens was surely a portent, such as the Scriptures record, of an approaching union of East and West—thus a good augury for the success of his enterprise.

That evening, towards the setting of the sun whose rising had moved him so deeply, Lesseps saw the Viceroy alone in his tent, and broached the subject of the Suez Canal. He had prepared a memorandum, on the basis of which he now addressed him, avoiding at this stage too much detail, technical, financial or political, and confining himself to the broad facts and arguments. He sketched in the ancient history of the Canal, then turned to

Napoleon's 'grand idea' of restoring it. He quoted a remark of Napoleon to Le Père, that it would not now be his own fate to carry it out, but that one day a Turkish Government would achieve glory by doing so.

The time had surely now come, Lesseps suggested, to realize this prediction. To those who saw only decadence and ruin in the Ottoman Empire, the piercing of the Isthmus would prove that it was still fruitful and capable of adding a brilliant page to the history of the civilized world. That Empire's importance, in the eyes of the world, lay in its possession of the Bosporus, the channel between the Black Sea and the Mediterranean. A channel of similar importance, across the Isthmus of Suez, would place it in an unassailable position, for the European powers would be obliged, for the sake of their mutual security, to guarantee its neutrality.

Lesseps briefly summarized the work of prospection done not only by Le Père but more recently by Linant and Talabot, and by two French engineers, Gallice and Mougel, in Mohammed Ali's service. All had agreed that, under modern engineering conditions, the Canal could be made. It remained only to work out the cost, which, however high, could hardly be out of proportion to the ultimate profit and benefit of so great an enterprise. He had drawn up a table of comparative mileage to prove that the Canal would reduce at least by half the present distance between the countries of Europe and India. He gave an estimate too of the tonnage of shipping liable to pass through it, in relation to that now passing around the Cape, and of the consequent profit to world trade in general and to the revenue of Egypt in particular.

In terms of grandeur, as well as of practical utility, there could exist no parallel to such a work as this. Lesseps knew well how to strike the imagination and flatter the vanity of his illustrious patron:

For his reign, what a fine claim to glory! For Egypt, what an imperishable source of riches! The names of those Egyptian sovereigns who built the Pyramids, those monuments to human pride, are forgotten. The name of the Prince who opens the great maritime canal will be blessed from century to century until the end of time.

Setting the Canal in a religious perspective, Lesseps stressed its advantages to the pilgrimage to Mecca and to Moslems throughout the world. Setting it in a global perspective, he compared unfavourably the projected canal across the Isthmus of Panama, which owing to the mountainous terrain would have to be broken by stretches of railway, with a canal across the Isthmus of Suez, which could run unbroken from one sea to the other, and thus complete at less cost and inconvenience a link quite as valuable to the world's great lines of communication. Its construction, so important to posterity, could no longer be seriously opposed. On the contrary, it could only arouse universal sympathy and the active co-operation of enlightened men in all countries.

Lesseps, in expounding his project, combined imaginative vision with rational, clear-headed argument, and a force of expression inspired by conviction. Mohammed Said listened intently throughout. Then he asked him a number of intelligent questions, which led to a two-hour discussion between them. Finally he said: 'I am convinced. I accept your plan. During the rest of our journey we shall consider ways and means of carrying it out. It is an understood thing. You can count on me.'

The Viceroy then summoned his generals, who sat around him on camp stools, while he retailed the conversation he had had with 'his friend', and invited them to express their opinions. Little able to comprehend such an enterprise, they gazed wide-eyed at Lesseps, thinking rather of an equestrian feat he had performed that day, jumping a high stone wall in their presence with ease. So skilled a horseman, they reflected, could surely give nothing but good advice to their master, and as he spoke they raised their hands to their foreheads in a gesture of respect.

That night Lesseps drew up for the Viceroy a revised memorandum, on the basis of the information he had given him. On arrival in Cairo he lost little time in clinching and giving form to the agreement. He went straight to see Linant, the engineer, who had inspired his interest twenty years back. He told him the good news that the canal, of which Linant had dreamed for so long, was at last to be dug, and his old friend clasped him in an emotional embrace. Already he had recommended Linant to the

Viceroy, together with Mougel, as the engineer most suited to carry out a preparatory survey of the Isthmus.

Before leaving, he went upstairs and greeted Madame Linant, his host's Syrian wife. Since Lesseps, in his official Consular capacity, had married the pair of them, it was a happy reunion. Then he proceeded to the palace in which the Viceroy had allotted him quarters. By an appropriate coincidence it was the building which had, half a century earlier, housed Napoleon's *Institut d'Egypte*, and his commission of experts had met in its rooms to discuss for the first time the piercing of the Isthmus of Suez.

Next morning the Viceroy held a reception in the Citadel, where the officials and foreign consuls came to pay him their respects on his return to Cairo. To Lesseps's surprise and gratification, he announced publicly his intention to open up the Isthmus of Suez by a maritime canal. He proposed to charge Ferdinand de Lesseps with the task of forming a company of the capitalists of all nations, to which he would concede the right to construct and exploit it. The British Consul-General looked a trifle embarrassed. To the United States Consul-General, Mohammed Said jested: 'We are going into competition with the isthmus of Panama and we shall have finished before you.'

The Concession was signed five days later, on November 30, 1854. The Company which Lesseps was empowered to form for the purpose of piercing the Isthmus of Suez and exploiting a canal between the two seas, would be known as the *Compagnie Universelle du Canal Maritime de Suez*. Its President was to be named by the Egyptian Government, preferably from among the shareholders most concerned in the enterprise. The Concession was to run for ninety-nine years from the date of the opening of the Canal, after which it would revert, on payment of an agreed indemnity, to the Egyptian Government. The Company would construct the Canal at its own cost, and would be granted such public lands as it required for this purpose, together with the tax-free privilege of working mines and quarries, and of importing machinery and materials required. This covered lands for the further construction of a Sweet Water Canal, if required, linking the Nile with the Isthmus to provide it with a supply of fresh water, or for an indirect route for the maritime canal itself, on condition that the Company would, at its own charge, put the lands under cultivation.

Any fortifications required by the Egyptian Government would be a charge on the Company. The Government would receive 15 per cent of the annual net profits of the Company, in addition to dividends from shares it might purchase. Of the remaining profits, 75 per cent would go to the Company and 10 per cent to the founder shareholders, whose names were to be approved by the Viceroy. Drawn from among Lesseps's friends and associates, including the Canal engineers, their contributions were designed largely to cover the Company's initial expenses, and enable work to begin before its public formation. The tolls or charges for the use of the Canal, to be agreed between the Company and the Viceroy, would be the same for all nations. The Viceroy promised his loyal support and that of his Government officials, for the execution and exploitation of the work.

From now onwards Mohammed Said referred to the Canal as '*mon affaire*'. To Lesseps he said: 'It is for you to fix the sum I shall invest.' He agreed to the appointment of Linant and Mougel to carry out the final survey, though he had reservations as to whether they would agree with one another. He would have preferred to send only Linant, who had after all done the job already, verifying and amplifying his own original survey and that of Talabot in a further conclusive report a bare two years before. But Lesseps thought it prudent to have two opinions. Later the Viceroy reproached Lesseps for advancing money through his own banker to Linant, for the trip to the Isthmus, insisting that, as his guest, he was to undergo no expense.

With Linant and Mougel, Lesseps left for Suez just before Christmas, as Napoleon had done fifty-six years before. Sabatier, the French Consul-General, and his wife, accompanied them. The desert road was now macadamized. But the aspect of Suez was as wretched and its water as brackish as ever. 'Our Canal', Lesseps predicted, 'will give it the water and the movement it lacks.'

Like Napoleon—and perhaps Moses—he left with his party for an expedition to the Wells of Moses, on Sinai. Though the journey was made this time by steamboat, a gale prevented them from landing and consuming the roast sheep prepared for them. They camped, with twenty barrels of Nile water for their trip through the Isthmus, at the point where the Nile-water canal, planned by Linant as an auxiliary to the maritime canal, would eventually emerge. They read Linant's previous memoranda, and Lesseps read his Bible as they approached the land where Joseph and his people had settled and whence Moses rescued their descendants from servitude four centuries later. It was the last day of the year 1854, and as Lesseps left his tent at sunset, he saw the last flash of a meteor which had shot across the sky from East to West. Illuminating the start of their journey of prospection, could this be another favourable omen?

A fortnight later they returned to Cairo. Linant and Mougel had agreed on the practicability of the Canal, and Lesseps gave them a list of nineteen points on which they would base their definitive report for the Viceroy. Looking up at the Cairo Citadel, Lesseps thought of the military *coup d'état* which, with his father's encouragement, had brought Mohammed Ali to power. Now, with his own encouragement, Mohammed Ali's son had staged the pacific *coup d'état* which would complete the rehabilitation of Egypt. No country would profit more from this *coup* than Britain. But would Britain realize it?

Lesseps, pondering on the scheme in his journal, feared not. It struck too deeply at Britain's 'old egotistical policy'. Since his youth he had been following the policy of the British in Egypt at every stage. Why had they used their power to defeat Napoleon's expedition? Why, later, had they protected the Mamluks, who divided and held back the country? Why, in 1840, had they united all Europe against France and Mohammed Ali? Why had they supported Abbas, the fanatical enemy of progress?

Lesseps's fears, however valid his interpretation, were to be abundantly justified. The Suez Canal had at last been officially sanctioned. But only now did its real troubles begin. During the years that followed, the British Government, consistently, emphatically and implacably, was to make every conceivable move to prevent a Suez Canal from being made.

CHAPTER V

Obstacles at Constantinople

The agent of British policy in Egypt was now Frederick Bruce, Consul-General in Alexandria. Son of an old Scottish family, his father, the seventh Lord Elgin (famed for his rescue of the Elgin marbles), had been Ambassador to the Porte at the time of the British expedition against Napoleon in Egypt, while his elder brother was to become Viceroy of India.

In anticipation of trouble, Lesseps had taken the precaution of calling upon Bruce to give him the news of the Concession in advance of its public announcement. Bruce, unable yet to speak for his Government, was guarded in his initial reactions. But, expressing his personal opinion to Lesseps, he did not seem to anticipate opposition from home, provided the Canal remained a private commercial proposition, free from Government intervention or influence by any one power. Lesseps followed up the interview with a persuasive letter to Bruce, stressing the value to Britain of this 'work of civilization and progress', and its potential benefits to the Anglo-French alliance. This alliance, surviving previous vicissitudes, had now been reaffirmed in the Crimean War against Russia, with the rise to power of the Prince-President as the Emperor Napoleon III.

But, on second thoughts, Bruce's attitude hardened. Acting still 'in the absence of any instructions' but with a shrewd idea of his Government's probable policy, he obtained an audience with the Viceroy. Finding him 'dazzled with the grandiose nature of the scheme', he sought tactfully to cool his ardour. The Viceroy should not commit himself prematurely to any such undertaking. It would impose too great a financial burden on Egypt's resources, at this time when the debts of Abbas were still undischarged, and its practicability and prospects could only be tested in the money-markets of Europe. Bruce stressed by contrast the financial advantage to Egypt of the railway, which had now almost reached Cairo from Alexandria and should be immediately extended to Suez.

Expressing his own views to London, Bruce considered that 'a direct Canal between Suez and the Mediterranean would give Egypt the go-by, would have a constant tendency to escape from the jurisdiction of the Egyptian Government, and would in no way enrich it except in so far as it might create a demand for supplies'. An indirect Canal would profit the country more, but it must be judged by the British and Turkish Governments how far the consequent presence within it of a powerful foreign company would affect its neutrality. These were in any case matters for consideration not by the Egyptian Government but by the Porte, whose assent to the scheme would be necessary.

Bruce's views quickly became known—and exaggerated—by his Consular colleagues. Soon after the announcement of the Concession, his predecessor, Murray, chanced to pass through Alexandria on his way to a new post as Minister in Persia. Known to be an enemy of the Canal—and reputed to be an enemy of France—he was assumed, in the small gossipy world of the Consuls, to have influenced Bruce against it. Sabatier, the French Consul-General, reported to Paris that Bruce was strongly opposing the Concession among the Viceroy's entourage, and urging them to open their master's eyes to its perils. This, he had warned them, was 'nothing less than a matter of life or death' for Said. Britain would never consent to a Canal, and her Ambassador, Lord Stratford de Redcliffe, 'Sovereign Master at Constantinople', would know very well how to force the Turks to refuse their sanction. Bruce was even said by Sabatier to have threatened the Viceroy: 'Take care, the alliance between France and England hangs only by a thread, and it is a thread which you risk breaking if you persist in your project for a Canal. I still hope you will draw back before so great a responsibility.'

When this was brought to the notice of the Foreign Office, Bruce denied that he had used any such words to the Viceroy—

and they may indeed have been put into his mouth, in a spirit of mischief, by Said himself. Lord Cowley, the British Ambassador in Paris, reassured the Quai d'Orsay accordingly. Britain's opposition was due to no spirit of rivalry with France, and would not be pushed 'beyond the bounds of courtesy'. All that Bruce had done was to throw cold water on 'the scheme in its present shape'.

By now Bruce had received his official instructions from London, which he had anticipated correctly enough. He should represent to the Viceroy, 'in a friendly manner', Her Majesty's Government's opinion that the scheme was in the first place impracticable, and in the second place called for too large an expenditure. This Egypt would be unable to meet, and if it was met from elsewhere the necessary conditions would cripple the Viceroy's freedom of action in administering the country. The fact that the proposal came from a Frenchman was quite immaterial. 'Whatever tends to facilitate intercourse between the British dominions in Europe and in Asia must necessarily be agreeable to Her Majesty's Government', whether proposed by a native of France or of Britain. But in the Government's view, this scheme would produce no such advantage, and on the other hand might well be injurious to the interests of Egypt. Thus Bruce was instructed to keep aloof from it and to state frankly to the Viceroy his reasons for doing so.

The French Government, as anxious as Britain to preserve good relations between the two countries, was outwardly almost as non-committal. To Lord Cowley, Drouyn de Lhuys, the Foreign Minister, positively disclaimed any official interest in Lesseps's plan, and his Consul-General was instructed in this sense. On the other hand, the Minister gave his opinion that the scheme, if carried out, could benefit Egypt and mercantile interests in general. Thus his Government, 'without interfering at all would yet see it executed with satisfaction'.

Meanwhile the present key to its execution lay not in Egypt, not even in Paris or London, but, as Bruce had remarked, in Constantinople. The Viceroy, although since 1841 he had enjoyed wide prerogatives, was still the Sultan's vassal, and Said

considered it impolitic to regard as within his exclusive independent authority so major a work as the construction of a Suez
Canal. He thus informed the Sultan of his Concession, putting it
forward, like the railway project, as a means of relieving, through
European capital, Egypt's present financial embarrassment. A
codicil was added to the Concession, providing for the Sultan's
prior ratification.

The Viceroy regarded this less as a duty than as a gesture of
courtesy, but deferred a start on the work pending the Porte's
consent. The Porte however delayed its reply, pending guidance
from its representatives in Paris and London. Thus, on Lesseps's
return from the Isthmus, the Viceroy sent him to Constantinople
with a personal note to the Grand Vizier, Reshid Pasha, in the
confident hope that he would be able to hasten the authorization.
Lesseps, his natural optimism enhanced by a touch of his former
political innocence, was confident too. The Sultan could hardly
respond with a veto to his Viceroy's sincere and deferent gesture.
The British, for all their objections to a Canal, could hardly
obstruct the French in an enterprise of such benefit to all the
world, at this moment when their two armies were fighting side
by side in the Crimea, in Turkey's defence. The Turks might
hesitate, but they surely would not refuse.

Lesseps was to be disillusioned. He had underrated the full
extent of the power of Lord Stratford de Redcliffe, the British
Ambassador. Stratford Canning, lately raised to the peerage, had
spent most of his career in Turkey, and had 'reigned' over her, as
Ambassador—with a single short break—for twelve years past.
His majesty and authority were such that he was known in
Constantinople as Sultan Stratford, or Abd-ul-Canning. Gracious
in manners, cultivated in mind, he would converse with charm
until a political difference arose. Then, as Persigny, later French
Ambassador in London, once said of him, 'Immediately you
heard the roar of the British lion'. He personified British policy
in his determination to hold the Ottoman Empire together by
means of internal reform. The Turkish Ministers feared and
revered him, more as a monarch than as a diplomat, and he was a
master of the forces of 'intrigue by night and intimidation by day'.

On the subject of the Suez Canal he had not yet received instructions from London. But this did not incommode him. Lord Stratford was used to acting on his own responsibility, and months would go by in which he would follow an independent course of action, with the implicit acquiescence of London. On the Canal his immediate views were clear. He had been a consistent sponsor of the Euphrates route. With regard to Suez, he had sponsored the railway at the Porte, and he now advocated to London its rapid completion. The French, in opposing the plan for the railway, had insisted, despite the British, that the Viceroy refer it to the Porte for sanction, and this Stratford had been able to secure. Now it suited the British to insist, in their turn, that he refer to it the French plan for the Canal, and this he was resolved to defeat. In general Stratford mistrusted the French, seeing in them a threat to his policy of conserving the Ottoman Empire and suspecting them of a self-interested resolve, now as in 1840, to detach Egypt from the Sultan's dominions.

Indoctrinating the Grand Vizier, Reshid Pasha, he thus urged him to 'discountenance the plan', and this he undertook to do, on the grounds that the time was inopportune. The result was a letter from the Grand Vizier to the Viceroy, favouring the completion of the railway, but temporizing on the Canal with a request for further detailed information. Reshid was an intelligent Circassian, who had spent much time in France and Britain, and aspired seriously to introduce Western reforms into Turkey. Stratford, who had known him since his youth, wielded a strong influence over him and was confident that he could rely on him to carry out his wishes.

Lesseps, on his arrival in Constantinople, soon learned of Lord Stratford's hostility. Received by Reshid, he resorted to the argument that the Ambassador's views were personal and did not necessarily reflect those of his Government. He deprecated to the Grand Vizier, as harmful to the Sultan's dignity, the domineering conduct of this foreign agent, with his capricious discontents. By contrast he stressed the loyal friendship of an enlightened Prince, the 'right arm' of the Empire as the Sultan was its head, whose pride was publicly engaged in the scheme for a Canal and who was now making to his sovereign an act of deference concerning it. No power, he argued, could threaten the integrity of Turkey through a Suez Canal. For its neutrality, and that of Egypt, would be guaranteed internationally as that of the Dardanelles had been. None could contest the right of the Sultan to act on his own. But if he required the support of any one power, he could count upon that of the Emperor Napoleon III. So Lesseps was bold enough, without authorization, to assert. He was backed up discreetly by the French Ambassador, Benedetti, similarly taking his sovereign's name in vain when the occasion seemed to demand it.

Lesseps found Reshid non-committal. Nevertheless his arguments made their mark. The Question had been referred to the Divan, the Sultan's Council. Its ministers were inclined, at first, to confirm the Concession, reluctant to displease, by a refusal, both the Viceroy and the Emperor. The French had regained some prestige with the Porte as a result of the Crimean War, and British influence was no longer unchallenged as it had been since 1840. Lord Stratford, used as he was to acting on his own authority, now found himself, thanks to Lesseps, in an unfamiliar dilemma. This, he confessed to Lord Clarendon, the Foreign Secretary, was 'not a little embarrassing', and he pressed hard for official instructions lest, in default of British objections, the Porte confirm the Viceroy's grant. The Viceroy, after all, had done his duty in applying for sanction, and 'I know not as yet by what means, if any, I shall be able to meet Your Lordship's wishes as to the Canal without incurring a serious hazard.'

Meanwhile Stratford made the most of His Lordship's instructions to Bruce, which he had seen, using them to frighten Reshid off 'giving a rash assent to so expensive, ill-timed, and questionable an enterprise'. To gain time, he made known to the Divan indirectly, through go-betweens, his view that this was not a matter for immediate decision. It was too complex and vast 'to be decided offhand between a private individual, however respectable, and a provincial Governor, however talented, having but little administrative experience'. The Divan thus postponed a decision.

Lesseps, disturbed at this rebuff, resolved to tackle Lord Stratford in person. Taking advantage of an invitation to dinner *en famille* at the British Embassy, he launched upon the Ambassador 'an oral assault', as Stratford later described it to Bruce, training his batteries in such a way as 'might have saved General Canrobert the trouble of taking Sevastopol'. Lesseps argued that it would be useless to resist a project in itself so useful and so powerfully supported. Lord Stratford complimented him on his frankness. He declared that he himself had no preconceived views on the Canal. It was a grand conception; if Lesseps succeeded in carrying it out, it would do him great honour. 'But', he added, 'it will only be realisable in a hundred years' time. The moment is inopportune.'

To this Lesseps replied: 'My Lord, if the affair is inopportune for you, who do not wish it, it is opportune for me, who wish it, and since you admit that it would be useful and will do me honour, why put it off for a hundred years? Since by that time I shall be unable to see the Canal made and since I have a complete faith in the possibility of its early execution, I am in a hurry to enjoy it. You yourself should be in even more of a hurry than I am.'

Next day Lesseps sent him a memorandum. In this he ventured to hope that it would never be said of Britain, fighting against Russia for civilization, the freedom of the seas, the independence of Europe and the integrity of Turkey, that she alone opposed an enterprise so favourable to the alliance between the powers and to the pacification of the East. His talk with the Ambassador had renewed many of his misgivings, and he proposed a further rendezvous for the following day. But Lord Stratford was not to be drawn. He evaded the rendezvous. This was a project, he replied, which closely concerned the interests of more than one country and, while attractive in theory, caused differences of opinion in practice. It was a matter of high policy, on which personal views must give way to official contingencies.

Undaunted, Lesseps wrote to him again, launching into flights of idealism and citations of the workings of Providence. Extolling the benefits of the Anglo-French alliance, he argued that Egypt was the only land on the face of the globe over which the two allies might still disagree, as they had done in 1840. Now a providential event promised to remove this source of conflict. For the geography of the globe could be changed. Human industry could pierce the Isthmus of Suez. The route to the East would pass along the boundary of Egypt, becoming accessible to all the world, and no longer through the heart of the land itself, which would thus cease to be a bone of contention between any one nation and another. Thus the union of the two seas would conserve as indissoluble the union between Britain and France.

But Lord Stratford de Redcliffe was impervious to such visionary arguments. The Divan was now to meet once again, and he was making sure, in a long interview with the Grand Vizier, that it would again postpone a decision. Once again he succeeded. This time the Ministers decided to refer the Canal project to a special Commission. Rather than compromise his master by acceding to this, Lesseps returned to Cairo, empty-handed but for a letter from the Grand Vizier to the Viceroy, assuring him that the Canal was still under study by the Council of Ministers, but at least admitting it to be a most useful enterprise. Lord Stratford 'felt a relief when he went away', though doubtless 'with the intention of returning ere long to achieve his triumph'.

*   *   *

Lesseps's journey, if disappointing, had not been useless. He had not foreseen that the British would prove so active, or the Turks so passive. He was at least now familiar with the pattern of diplomacy in the declining Ottoman Empire, which was to enmesh the promoters of the Suez Canal for a decade to come. It was a negative pattern in which procrastination, as a weapon of policy, prevailed over action; a pattern of intrigue and manipulation, oscillating now to one side and now to the other, in the hands of ministers anxious not to displease yet not over-anxious to please, who thus ended in favouring nobody and achieving few if any positive results.

In this devious struggle for influence, Britain still prevailed over France. Lord Stratford had won the first skirmish in his campaign against the Suez Canal. But now the tactical balance was swiftly to change. For his friends at the Porte overplayed their hand, and provoked a crisis at the Court of the Viceroy of Egypt. From Constantinople Said received two notes, both evidently inspired by a single mind. One came from his brother-in-law, Kiamil Pasha, President of the Divan; the other from the Grand Vizier himself. Both were designed to open his eyes to the perils facing himself and his country if he were to persist in this 'fatal project'.

The first raised questions concerning the Canal, on the guarantees to be secured from the Company regarding the passage of warships; on the concessions of lands to Europeans, which seemed contrary to Turkish practice. And it warned Said against embroilment in 'a certain struggle with England'. To these objections Lesseps, now back at his master's side, had the answers ready: following international precedent, no alienation of Turkish sovereignty would arise; the concessions to Europeans applied only to uncultivated lands, whose irrigation would be a profit to Egypt. Nor was it contrary to Turkish practice to make concessions to an international as opposed to a national company. The Canal could only benefit Britain, whose hostility was not to be feared; nor had the British Consul-General made any official representation concerning it on behalf of his Government.

The second letter, from Reshid Pasha, was phrased in more outspoken and indeed in menacing terms. He regretted that His Highness had thrown himself, as his father had done before him, into the arms of France, whose Government had no more stability than its agents. 'France', he wrote, 'can do nothing for you or against you, while England can do you a great deal of harm from the moment you lose her support. The agents of the English are always sustained and upheld. There is therefore no real way of dealing with them, the less so because they never forget their grudges.' The Sultan was greatly irritated against his Viceroy, and could only be appeased if he would cease to speak of the Canal. Otherwise he ran the risk of seeing a British naval squadron from Sebastopol before the port of Alexandria.

This attempt at intimidation had the opposite effect. It at once stiffened Said, who had fallen into a mood of discouragement at the Porte's aloofness. To Sabatier, the French Consul-General, he exploded: 'Reshid Pasha is only a comedian and a perfidious and corrupt intriguer. He is mistaken if he thinks he can exploit me as he exploited my father. I will not be his dupe.' As an instance of his personal perfidy he confided to Lesseps that Reshid had sent him a present of a beautiful slave-girl, asking afterwards whether she had satisfied him and whether his wife was displeased. Said, so he declared, had not even looked at her. His reply to the Grand Vizier was simple: 'I am neither a Frenchman nor an Englishman. I am and I shall remain an Egyptian Turk.' As for the Sultan, the Viceroy trusted that 'instead of depriving me of his good graces, he will support me against my enemies, who are also his'.

With Kiamil Pasha he threatened privately to break off all personal relations, if he continued to act and write in this sense. Replying to him formally, on Lesseps's brief, he combated the accusation of a pro-French bias with the retort that 'you are all of you in great fear and mortal terror of Lord Stratford de Redcliffe and the English'. In regard to the making of the Canal, he had behaved like 'a good Turk'. The enterprise depended on no single power but on 'a general company composed of people from all countries'. Nor was it the British Government that raised the objections, but 'the said Canning alone who does it in the interest of his own personal policy'.

To Sabatier the Viceroy launched into a long disquisition on the benefits of the Canal, 'with a fervour and a precision of mind', wrote the Consul, 'of which I have seen no comparable example in any Turk.' The utility of the Canal, Said insisted, was so indisputable that, 'if I did not make it today, Europe would impose it on me later. She would have to be blind, as they are at Constantinople, not to see that at first glance.'

Thus the manœuvre rebounded against its authors. Benedetti, the French Ambassador, as skilled as Lord Stratford in the art of diplomatic intrigue and less scrupulous in the use of its more dubious crafts, came into possession of the two letters. These gave him the chance to protest to the Grand Vizier against breaches of diplomatic propriety in his relations with France, and his protest was reinforced from Paris. Reshid was obliged to resign, and his example was followed by Kiamil—who afterwards wrote to his wife in Egypt that Reshid was 'up to the beard in the waters of Lord Stratford'.

Here was a set-back for the British Ambassador, who chanced to be away in the Crimea at the time, and a success for the French. Benedetti now came out in more open and active support for Lesseps, going so far as to claim that the Sultan should authorize the Canal, on the grounds of the Viceroy's reply to his questions— which had now been received—and the continued absence of official objections from Britain.

Thus, for the moment, Lesseps need fear no more trouble from Constantinople. 'Here, as a start to our voyage,' he wrote to his brother in Paris, 'is a statesman in the sea; and perhaps he will be followed by others.' The centre of gravity shifted to Europe, where London now took up the Canal question in diplomatic exchanges with Paris.

In Egypt the engineers, Linant and Mougel, had completed their preliminary report, reaching the expected conclusion that the Canal, with a port at either end of the Isthmus and a third inland, could be made. This was to be translated into all languages, and would then be submitted to a Scientific Commission of Enquiry, composed of Europe's leading engineers. The Commission's conclusions would serve as a basis for the organization of an international company to carry out the Concession.

The time was now opportune for Lesseps to leave for France. Early in June 1855 he arrived in Paris to launch on his home ground a propaganda campaign for the Suez Canal. This he was to spread, systematically and thoroughly, to all corners of Europe.

*   *   *

Lesseps shrank from plunging back into the 'wasps' nest' of European diplomacy. Had it been possible, he would have preferred to put the horse before the cart, as he saw it, by first starting on his scheme for the Canal, and then submitting it to the powers for ratification. 'I always feel very strong on my own Egyptian ground', he wrote to his mother-in-law, 'and . . . I shall persevere in refusing to believe that the answer is either in Paris or in London. London will say "No", Paris "Yes". Who can guarantee to me that the Yes will be stronger than the No?'

In Paris he placed his hopes largely upon the Emperor Napoleon III in person, and on his wife Eugénie, who had become Empress two years earlier. A daughter of the noble Spanish family of Montijo, she was Lesseps's cousin, to whom he had done a good turn during the troubles in Madrid, and whom he liked to claim as his august patroness and 'guardian angel'. For her part she recognized in him a kindred spirit, seduced by the grandeur of his conceptions, treating him as a friend, and now letting him know that the Emperor would make the Canal project his own. 'The thing', he had assured her, 'shall be done.'

But not, it seemed, yet. In an audience with the Emperor, Lesseps found him hesitant—for the first time but not for the last. He showed a benevolent interest in the scheme, and hoped Lesseps would persist in his policy of proceeding unofficially, without reference to Governments. When Lesseps put forward a Quai d'Orsay proposal that both the French and British Governments should abstain from intervention, leaving the matter to be settled between Turkey and Egypt, the Emperor gave a vague answer and in general showed an alarmist attitude—no doubt induced by reports from his Ambassador in London. 'If I supported you now,' he went so far as to say, 'it would be war with England.' But he added that when the interests of European, and particularly French capital were engaged, 'all the world will support you, and I shall be the first to do so'. From this negative reaction it became clear that Lord Palmerston's positive views on the subject had produced their intended effect on the Tuileries.

At the Foreign Office they had been made clear enough in Palmerston's minutes on Lord Stratford's despatches. On one he had written:

If this Canal is simply to be a barge passage for boats and vessels not fit for the seas, it will afford no material advantage to commerce superior to that which is afforded or will be afforded by a railway from Alexandria to Cairo, or from Cairo to Suez; and the expense of making a Canal would be very great, because besides the Canal itself, which would be liable to fill up with blowing sand, a harbour would be required at each end, and from the shallowness of the water for two or three miles on each side such harbours would be difficult to make. . . . If however the Canal is meant to be one for seagoing ships, the expense would be enormous and the undertaking would never pay. But it would be injurious to England because, in any quarrel between England and France, France, being so much nearer to the Canal, would have much the start of us in sending ships and troops to the Indian seas.

On another he wrote:

It is quite clear that this scheme is founded on ulterior motives hostile to British views and interests and the secret intention no doubt is to lay a foundation for a future severance of Egypt from Turkey and for placing it under French protection. A deep and wide canal interposed between Egypt and Syria studded with fortifications could be a military defensive line which, with the desert in front of it, would render the task of a Turkish army very difficult. . . . If land is to be conceded to the French company, a French colony on French territory would be interposed between Turkey and Egypt and any attempt by Turkish troops to cross that line would be held to be an invasion of France. . . . It seems to me that these considerations might be frankly and unreservedly explained to the French Government and they might be asked whether they think it worth while to endanger the alliance by pressing forward this scheme.

These forthright observations now crystallized into an official despatch from Lord Clarendon, the Foreign Secretary, to Lord Cowley, the British Ambassador in Paris. Here was the first concrete statement of British policy, and its communication to the Quai d'Orsay coincided with Lesseps's arrival. It embodied three main objections to the Canal. First, its construction was, for specified reasons, physically impossible, except at a cost which would rule it out as a profitable commercial speculation. Hence the motives for the scheme could only be political. Secondly, it would delay, if not prevent, the completion of the railway through Egypt, to the detriment of Britain's Indian interests. Britain required no ascendancy, no territory in Egypt. But she must have a free, rapid, and secure thoroughfare to India, and this the railway would furnish as long as Egypt remained a dependancy of the Ottoman Empire.

Lastly, the Canal scheme was founded on an antagonistic policy on the part of France, which the Government had hoped no longer prevailed—the policy of detaching Egypt from Turkey, to cut off Britain's easiest channel of communication with India. With this object, fortifications had been constructed by French engineers along the Mediterranean coast, and a great barrage on the Nile, to flood the Delta in case of military need. The Canal scheme was conceived in this military spirit, and should give way to the better policy which had succeeded it—that of friendly relations with Britain. Lord Stratford de Redcliffe was instructed in a similar sense.

As to the proposal that the two Governments should agree to interfere no further, but leave the matter to be decided between Sultan and Viceroy, this would simply 'throw back the question to be fought for locally by the cabals and intrigues of the agents and partisans of the two countries in Turkey and Egypt and would tend to revive and aggravate all the jealousies and rivalries which both Governments are now labouring to extinguish'. The proposal had emanated from Count Walewski, the new French Minister of Foreign Affairs, to whom it now fell to reply to these representations.

Walewski was a close confederate of the Emperor—and indeed a close relation, for he was a natural son of the great Napoleon himself, by the Polish Countess Marie Walewska. He was also an old acquaintance of Lesseps, and a warm partisan of the Canal project, which he had himself discussed with Mohammed Ali in the course of a mission to Egypt in 1840. He thus consulted Lesseps before drafting a long reply to Clarendon through the Comte de Persigny, another of the Emperor's 'swell mob' (as Clarendon described them) who had just succeeded him as Ambassador in London.

Setting aside for the moment his proposal for non-intervention, he proceeded to refute, one by one, the three British objections. First, if the construction of the Canal really proved to be technically impossible, its opponents had no cause for alarm. Similarly, if it proved to be financially impossible, international capital would fight shy of it. But the case seemed to be otherwise, from the favourable report, soon to be published, of the engineers, and their estimate that the total cost of construction would amount to 160 million francs—half that of the railway between Paris and Lyon. This report was to be submitted by the Viceroy to an international group of technicians for a final decision. Thus the Governments of Britain and France need not concern themselves with the scientific or financial aspects of a scheme free from political motives.

Secondly, the French Government had recently announced its support for the railway, though it was designed only to serve British interests, and here was surely a proof of French goodwill towards Britain. France sought even less than Britain any ascendancy or territory in Egypt. She believed in freedom of transit through Egypt over the shortest route between Britain and her Indian possessions. Nor was there any danger that resources devoted to the Canal would postpone the completion of the railway, since the Viceroy had decided to continue it, at his own expense, from Cairo to Suez, and preparatory work on it had already begun.

In reply to Clarendon's accusations of French antagonism towards Britain, Walewski concluded: 'If the Emperor's Government thought that the present Canal project, for which it has no responsibility, were inspired by a policy of antagonism, it would oppose it without hesitation. But this is not so.' On the contrary, the French Government was happy to see the present change in the relations between the two countries, which it was fostering sincerely. Nothing was more likely to revive the bad old sentiment of jealousy and mistrust than an unjustified opposition to an enterprise of world-wide importance. In the light of their new policy the two Governments should not allow themselves to be led astray by the prejudices of an earlier epoch, but should examine the present question with impartiality.

This despatch reflected advice from Lesseps. Now, with the blessing of the Emperor and an introduction from Walewski to Clarendon, he departed for London. The two adversaries, Lesseps and Palmerston, the irresistible force and the immovable mass, were now for the first time to confront one another in person.

CHAPTER VI

'The Man of 1840'

Lord Palmerston was now seventy years old. Born and reared in the eighteenth century, he had been in office, with brief interruptions, since the age of twenty-five. For fifteen years in all—between 1830 and 1851—he had ruled as Foreign Secretary. Now, in 1855, he had become Prime Minister.

Europe had changed in these seventy years, and was still changing. Palmerston himself had changed relatively little. Always wary of political abstractions, he still danced 'the perpetual quadrille of the Balance of Power', as a historian has called it.[1] He had learned its choreography from eighteenth-century mentors, illuminated by reason and enlightened self-interest, and in a nineteenth-century adaptation from immediate seniors like Canning. 'For Europe . . . read England'—so Canning had pronounced. British interests must predominate, and they depended upon European security.

This could best be achieved, not by alignment with such continental groupings as the absolutist Holy Alliance between Russia, Austria and Prussia, but by independent, judicious and decisive support for those democratic and nationalist forces which had risen to the surface of Europe since 1848. Palmerston sought, through the encouragement of nationalism, to prevent the domination of Europe by any imperialist power. But he still persisted in seeing a threat to his Balance of Power from Britain's hereditary enemy, France.

Palmerston had spent his boyhood and youth in the shadow of Napoleon's conquests in Europe. He had served as Secretary for War through his later campaigns, and toured his battlefields with the Duke of Wellington, who had only lately died. He had lived

[1] A. J. P. Taylor: *The Struggle for Mastery in Europe.*

through the dread of Napoleon's invasion of England; he had known fears of it since, when the Duke became convinced that Louis-Philippe planned another invasion, and when, just before the 'old warrior's' death in 1852, 'Young Nap upon a prancing horse' roused the alarm of a timid Government and a Francophobe public opinion; and he was to know fears of it again in 1859, when dark rumours of a specially-built French fleet of flat-bottomed boats aroused a new invasion scare, with panic defence measures on the Channel coast and demands for rearmament.

As Foreign Secretary Palmerston had instantly recognized Louis Napoleon's seizure of power as Emperor, seeing in it the only workable alternative to a divided Republic, which had in itself been an improvement on a discredited Anglophobe monarchy; and he had been forced to resign for his precipitate action. As to the Emperor himself Palmerston maintained reservations. Publicly he could praise him for his good faith, straightforwardness, and loyalty to the Crimean alliance, in a toast which he chose to define as 'entirely a new one since the days of the Crusaders'. Privately he would sum him up as 'an able, active, wary, counsel-keeping but ever-planning sovereign', with a mind which 'seems as full of schemes as a warren is full of rabbits'. Essentially France, which had had five revolutions in sixty years—two republics, two monarchies, and now an imperial dictatorship—was still seen as an unstable and unpredictable power. To combat Russia, she had involved Britain in an uneasy alliance to fight an unnecessary war; and now it was her evident design to bring her into an incomplete peace—or even to sign one without her—before the Russian threat had been finally and effectively disposed of.

An explanation of this attitude of Anglo-French distrust, if not an excuse for it, lay in the ambiguity of the Emperor's policy, and the frequent ineptitude of its tactics. Taciturn, lugubrious in aspect, secretive in manner and enigmatic in outlook, with an apparent absence of principles, he did not easily inspire confidence. Still less did his 'swell mob' of mendacious and unscrupulous Ministers. He was, after all, a military dictator, who had

seized power through an Army *coup d'état*; and he was a Bonaparte, steeped in the Napoleonic tradition, who might well be
tempted, as Lord Cowley saw it, to prove to the French people: 'I
have done for you what my uncle could not do.'

At the Tuileries Napoleon III reigned in an aura of martial
display. 'The Empire', he had declared on his accession, 'means
peace.' Yet, as he admitted, the settlement of 1815 still rankled
with France, and he was resolved to redress it, to remake the
map of Europe, with adjustments of what he took to be France's
legitimate frontiers. Inevitably this involved wars, or at the best
a state of armed peace with rumours of wars, which he fostered.
He aspired to become a power in Europe, as his uncle had been,
and as a means to this end struck belligerent attitudes.

None the less his end was more political than martial. His
Bonapartist ambitions for personal power and glory were counter-
balanced by dreams of a new order in Europe, a Utopian community, in which the free revolutionary nations would combine
peacefully in pursuit of international brotherhood. But in the
foreign policy of the Second Empire practice was too often at
variance with precept. In one aspect of it Napoleon was wholly
consistent—in his desire for a close liberal alliance with Britain.
Unfortunately, however, his actions and his methods of acting
too often inflamed long-standing British suspicions of France.
Statesmen other than those of the time, more disposed to foster
understanding and to lead rather than exploit public opinion,
might have eased or removed this mistrust.

Meanwhile the Emperor, for all his repudiation of territorial
aims, had an army of occupation in Rome; he was soon to attack
Austria without warning, and to grab Nice and—albeit with a
plebiscite—Savoy, for France. Within months of his *coup d'état*
he appeared to be threatening the independence of Belgium, a
move which was seen as a direct threat to Britain. But the most
serious obstacle to Anglo-French friendship lay in Napoleon's
supposed imperial intentions overseas, in the fact that, as
Palmerston saw it, 'both countries like two men in love with the
same woman' wished to predominate in the East. Just before
dubbing himself Emperor, Napoleon chose, in a speech at

Marseilles, to revive an idea of his uncle that the Mediterranean should become a 'French lake'. In reply to Lord Cowley's anxious enquiries the Foreign Minister explained with some embarrassment that the phrase was no more than a 'poetical image', designed to encourage the trade of Marseilles. But in the prevailing atmosphere, as Cowley advised, such ill-judged images could only increase foreign apprehension and justify press attacks from abroad.

In fact this Emperor without an Empire did not seriously covet one beyond the confines of Europe. His intervention against Russia in the Crimea, unlike that of Britain, had been prompted rather by European than by Eastern designs. But too often he looked like an imperialist. His *coup d'état* had been largely led by young officers from Algeria, on the southern shores of the 'French lake', whose colonial campaigns in the interior had stirred the Parisian imagination.

Later, reviving secret schemes of Louis Philippe, he showed signs of coveting Morocco, with a view, so Palmerston insisted, of fortifying the Straits of Gibraltar, and 'shutting us out of the Mediterranean'. In a spirit of appeasement, the Emperor broached a vague scheme for a partition of North Africa, in which France would have Morocco, Italy Tunis, and Britain Egypt. This had provoked Palmerston to the spirited comment: 'We do not want Egypt or wish it for ourselves any more than any rational man with an estate in the north of England and a residence in the south, would have wished to possess the inns on the north road. All he could want would have been that the inns should be well kept, always accessible, and furnishing him, when he came, with mutton chops and post-horses.' Now Palmerston smelt a new threat to this comfortable arrangement in the French scheme for a Suez Canal, threatening to command the Eastern end of the Mediterranean as the rock of Gibraltar commanded the Western. Such was the general political climate in which Ferdinand de Lesseps arrived in London towards the end of June 1855.

*   *   *

Soon after his arrival, he called on Lord Palmerston, who received him with courtesy. But he at once made it clear that his mind was made up and he was not to be shaken. Lesseps, in no way intimidated, broached his case with frankness, and begged him to reveal freely not merely his official but his personal objections. All he got from the Prime Minister, however, was a recital, word by word, of the familiar arguments contained in Lord Clarendon's despatch to Lord Cowley, which he had clearly dictated himself.

Lesseps returned the familiar replies with some elaboration of detail. But Palmerston disregarded them. Then, in a good-natured spirit he gave him an inkling of the real reasons which underlay his opposition. He confessed to two apprehensions: first, that Britain's commercial and maritime relations would be overthrown by the opening of a new route, open to all nations and thus depriving her of her present exclusive advantages; secondly, the uncertainty of France's future. For all his belief in the Emperor's sincerity and loyalty, he could not be sure of what might happen in France when he had gone.

Lord Palmerston was presiding with confidence over a world neatly and securely packaged within the bounds of nature's geography—on the Continent manipulating an intricate balance between the powers, in the Middle East upholding the *status quo* through the sovereignty of the Ottoman Empire. By this means he avoided costly commitments in relation to either, leaving Britain free to rule the oceans and develop an expanding commercial and political Empire. Now came the unwelcome revelation that geography itself was no longer immutable, thanks to scientific forces of which, as a man of an earlier age, he was ignorant, hence deeply mistrustful. To the consequent threat he reacted with alternate bursts of suspicion, indignation and disbelief.

Next day, at a soirée at Lady Palmerston's, Lesseps learnt from the French Ambassador, Persigny, that the Prime Minister had not been unfavourably impressed by their meeting. Lesseps knew none the less that Persigny was an enemy of the Canal. An early confederate of Louis Napoleon, first a non-commissioned officer in the Army, then a journalist in Paris, he had helped him, with devotion and daring, through the various stages of his rise to power, and had played an adventurous part in the final *coup d'état*. Napoleon himself once described him as crazy, and for all his vigour, he was often crass, stubborn, clumsy and deliberately truculent, with little of the tact and the subtlety required of a diplomat.

But Persigny was persuasive, and the Emperor, in his shifting and hesitant moods, responded to his personal influence, which was now strongly directed against support for the Suez Canal. Here he emerged as more British than the British, arguing that the Suez Canal would ruin Britain as the discovery of the Cape had ruined Venice. For it would subject her ships to the risk of flank attacks from Europe over a distance of thousands of miles, substituting for a safe and tranquil route a route full of danger.

For France this was a matter of future and secondary interest; for Britain, it was a matter of immediate and primary interest. To engage in a struggle for influence over a scheme so vague and uncertain would seriously prejudice, at this crucial moment, the Anglo-French alliance, of which Persigny had always been a passionate advocate. In this sense he spoke privately to his master, at the Tuileries, and his influence indubitably helped to discourage the Emperor from giving Lesseps the support of which his enterprise, on its merits, was worthy.

Clarendon received Lesseps as politely as Palmerston, but his opposition was even more emphatic. He declared that 'the objections of Her Majesty's Government were insuperable'. He enlarged luridly on the strategic threat of this 'great cut three hundred feet wide and twenty-eight feet deep . . . with fortifications on the banks of it and war steamers properly placed in it', and with both ends of it closed to Britain in the event of war with France; and he insisted that to assent to its construction would be 'a suicidal act on the part of England'.

Lesseps was not unduly discouraged by the British Government's attitude, which he had largely foreseen. Palmerston, he knew, was still 'the man of 1840', the man of the past. But Palmerston was not Britain—or so he believed. In this land of liberty there were countless men of spirit and intelligence who thought otherwise, and who, sooner or later, would carry public opinion. There were, for example, men like Richard Cobden. Cobden was the man of the Left, of the future, of the people whose 'voice . . . was the voice of God'. He was the apostle of that freedom of trade which would achieve universal peace. He was a strong supporter of the alliance with France, as an instrument for this peace; he respected the good intentions of Napoleon III, believing him to be 'capable of generous emotions' and, in 1860, seeking improved relations between the two countries, he was to persuade the Emperor to sign an Anglo-French commercial treaty. This was done in concert with Michel Chevalier, for whom Free Trade was almost a religious belief and who had been a fervent follower of St. Simon. Cobden was thus familiar with the St. Simonian pursuit of pacific ends through economic and industrial means, and it was natural that Lesseps should see in him a potential ally.

Immediately after obtaining his Concession, he had written to Cobden, 'as a friend of peace and of the Anglo-French alliance', telling him the news of his project and promising to give reality to the phrase '*Aperire terram gentibus*', which he now started to use as his motto. His letter, which summarized the essential benefits of the Canal, could if necessary serve, as he saw it, as the theme for a crusade against the men of the past. He expressed disbelief that his scheme could be opposed by any enlightened British statesman, and if it were, he trusted that public opinion, so powerful in Britain, would soon overcome self-interested and outdated objections.

But Cobden did not respond to his plea, nor to another two years later, partly through a reluctance to commit himself to so contentious a commercial and political venture as the Canal, and partly through a belief that the hazards of the Red Sea and of the Indian Ocean would discourage shipping from the use of the Suez route. This did not prevent Lesseps, in conversation back in Egypt, from claiming Cobden as a supporter, a claim provoking Clarendon to make known, through Bruce, that there was 'no public opinion in England favourable to Mr. Lesseps's projected canal, and that if there was Mr. Cobden is not in Parliament to express it'.

Meanwhile Lesseps had found one or two Members of Parliament to support him, and now started to lobby other sections of public opinion. First, at the direct request of the Emperor, he saw the editor of *The Times,* who had opposed Lord Palmerston. He received from him a promise not to oppose the Canal, and to advise accordingly his correspondent in Alexandria, who had been sending despatches unfavourable to it. The correspondent, none the less, continued to do so; but the editorial comment proved more favourable. Next, Lesseps tackled representatives of industrial, banking, shipping and commercial concerns, particularly those involved with Eastern markets, in a Britain where the Industrial Revolution was now in full tide.

Four years earlier the Great Exhibition had helped to open men's eyes to the new and glorious prospects of science and technology, and their role as a source of international progress. Thus Lesseps, with his sanguine disposition, was soon persuaded —or at least sought to persuade his associates—that British recalcitrance was a phantom which faded away as he confronted it. As a basis for his propaganda campaign, which from now onwards he was to wage with mounting zeal, he prepared and published, in English, a comprehensive pamphlet, designed to put before the British public, in detail and with lucid arguments, his plan to unite the two seas. This was circulated, with the relevant statistics and documents, to these various business concerns, to influential persons, to officials and Members of Parliament, and to the Press, with a request for their comments. He was able to announce to them that he had secured the services of James Rendel, a notable British engineer who had experience of canal and harbour construction, for the International Scientific Commission which was to pronounce on the scheme.

This pamphlet received some coverage, favourable and otherwise, in the London Press. The technical and financial aspects of the scheme were treated with a certain critical reserve. *The Economist*[1] saw no reason to believe that Lesseps's estimates were

[1] August 18, 1855.

'one sixpence more incorrect than the best of such estimates usually are', but remarked on their smallness—a mere six or seven millions—for such a huge undertaking: 'Why! It is not twice the cost of our inconvenient and gingerbread-looking Houses of Parliament'. Regarding the political aspect of relations between Europe and India, 'the time is entirely gone by when on this point that there can be any jealousies betwixt France and England'. There was no good reason for British alarm at the advantages given by such a Canal to the Mediterranean countries:

> Our naval power is so great that a facility in reaching India from England is a favourite object with us. It would add to our security. By our ships we could defend our possessions. The more other nations trade with England the more our possessions would be enriched. The more trade, too, we should have with them.

The *Athenaeum*[1] took an opposite view. Looking back into the past it posed the question, Would the advantages of the Canal in times of peace outweigh its dangers in times of war? Lesseps insisted on its neutrality, under a European guarantee— 'but when the French were chased from Toulon, and when they eluded Nelson, does anyone affect to believe that European guarantees would have prevented them from flying to the head of the Mediterranean, gliding into the Indian Seas, and committing havoc from Kurrachee to Malacca?' Suez would be a fortress, whose security would depend on the capacity of the Turks to defend it. If the defenders proved weak, 'that nation would command the passage which could forestall the others in seizing the position for itself'. The Suez Canal would be a new Dardanelles, a new Gibraltar, and 'English statesmen know that if such a fortress is to exist, it must be garrisoned by our forces, like Gibraltar'.

> At all events [the writer concluded], as the girdle of sea-sickness round these islands is valued as a cheap national defence, so the bank of land at Suez appears to a certain class

[1] August 25, 1855.

of politicians the barrier of our Indian empire—the breakwater which would keep the Indian waters tranquil, though a new Trafalgar had to be fought in the British seas.

The Edinburgh Review,[1] a periodical under Lord Palmerston's patronage, quoted Stephenson's verdict that the idea of a canal was impracticable. Even if it were practicable, it might benefit Egypt and Turkey, but 'would not touch the grand commerce of the world, which now follows the route round the Cape'. The Red Sea route might indeed be preferable for passenger and parcel traffic. 'But it is very questionable whether steamers will ever be able to compete with sailing vessels for goods traffic . . . At least nine-tenths of the cargoes of the world will continue for a long time to be carried by sailing vessels.'

Taking a glimpse into the future, the writer conceded that large steamers were now indeed being built, leviathans 'surpassing in size anything ever dreamed of'. Their size would be such as to put an end to all canals and to the project for the Suez Canal, as proposed, in particular. Thus 'the Red Sea may again be restored to its pristine solitude, undisturbed even by the weekly visit of the passing steamers'. English capital would inevitably find that the route round the Cape was 'infinitely preferable for commercial purposes'. This was a matter for decision, not by a mere international commission of engineers but by men thoroughly familiar with the commercial relations of the East and with conditions of navigation between Cape Gardafui and Gibraltar. Meanwhile 'the Suez Canal question may fairly be relegated among the *questions oiseuses* which may interest and amuse, but can hardly ever benefit mankind'.

*   *   *

The International Commission, the next logical step towards the completion of Lesseps's plans and their presentation to the world, assembled in Paris that autumn and appointed a delegation to proceed to Egypt. Mohammed Said received its members in a princely style befitting, as he put it, 'the crowned heads of

[1] January 1856.

science'. He entertained them to a sumptuous luncheon at which he presented his former tutor, and present private secretary, Koenig Bey, with the words: 'He has often put me on bread and water, but I do not serve them to him in return.' The chief members of the Commission were Conrad, its President, for Holland; for Austria, Negrelli, who had been a member of Enfantin's Study Group and knew the ground well; for Britain McClean, replacing Rendel, who was ill; and two Frenchmen, Renaud and Lieussoux. Collectively the delegation represented also Italy, Germany and Spain.

By the New Year of 1856 the delegation had covered the necessary ground on the Isthmus and completed its studies, using as a basis the findings of Linant and Mougel. In a preliminary report to the Viceroy—unanimous, but tailored to satisfy certain dissident views of McClean—it recommended that the sole method of joining the two seas was a direct canal from Suez to Pelusium. Its execution would be easy, its success and its benefit to the trade of the world was sure. The engineers rejected the indirect tracing to Alexandria on technical and financial grounds, but allowed for a branch between the main canal and the Nile, which McClean in particular favoured. The port of Suez offered a safe and broad anchorage, thirty-five feet deep. The port of Pelusium, if built eighteen miles west of the site previously chosen, offered an equally good anchorage, thirty feet deep. The cost of the construction was now estimated at 200 million francs (£8 million).

This and a final detailed report by the Scientific Commission gave Lesseps the ammunition and the added authority he needed to counter at least one of the arguments of his critics—that of the technical impracticability of the Canal. Thus armed, he proceeded to Paris, via Trieste, in time for the Peace Conference following the end of the Crimean War. This was the first European Congress for thirty-four years. Throughout this Spring, while it was in session, Paris became the hub of the Continent, thronged with Ministers, delegates, officials and persons of influence from all its countries. The corridors and *salons* of the Congress provided Lesseps with a fruitful lobbying-ground, in which to make contacts and to press his scheme in international circles, and for support he now had at his side engineers from the International Commission. With France now re-established as a respected power in Europe, in close alliance with Napoleon Bonaparte's former enemy Britain, the Congress chamber itself should, in theory at least, provide a suitable forum in which to launch upon the powers a scheme so conducive to international co-operation and peace as the Suez Canal.

Lesseps, though always sceptical of the channels of public diplomacy, saw no harm in testing theory with practice. With the encouragement of Metternich, whom he had visited on his way to Paris, he drafted a clause, covering the neutralization of the Suez Canal, for insertion in the Peace Treaty. Walewski approved it; Count Buol, the Austrian delegate, undertook, with the approval of his Government, to submit it to the Congress, and was assured of support from Russia, Prussia and Sardinia. But Britain, in the person of Lord Clarendon, flatly opposed it, suspecting veiled designs by France on Egypt; and it was clear that Turkey, jealous of her authority over Egypt, would do the same. Thus Buol, rather than commit either Government to an inevitable veto, dropped his proposal. No reference to the Suez Canal therefore appeared in the Peace Treaty, which was signed with the quill of an eagle at the end of March 1856.

The Emperor, after a banquet which he gave for the pleni-potentiaries, had a talk on the subject of the Canal first with Aali Pasha, the Turkish Grand Vizier, and then with Lord Clarendon, the British Foreign Secretary. In reply to Aali, who broached the topic on behalf of his master the Sultan, the Emperor avowed his sympathy with the scheme, as a benefit to all the world. He had, he assured him, studied it in detail; he disagreed with the objections to it, mostly from Britain, and he hoped to dispose of them. But for the moment he was not anxious to press the matter too far, for fear of disturbing the good relations between the two countries. Aali Pasha replied that the Porte favoured the scheme, subject to minor qualifications and to safeguards for Turkish suzerainty, as a source of profit to Egypt in which Turkey hoped to share.

The Emperor then summoned Lord Clarendon and asked for

his views. Taken a little by surprise, the Foreign Secretary was non-commital, saying that this was a matter for his Cabinet, but adding that he believed the Canal to be impracticable. The Emperor contested this, quoting scientific opinion, and pressed him further. Clarendon then disclaimed any commercial objections to the Canal—such as Palmerston had previously voiced to Lesseps. But he remarked that it raised delicate problems regarding Turkey's relations with Egypt, since the Viceroy could only construct it with the Sultan's permission. Clarendon repeated this view in an interview with Lesseps.

Meanwhile the Grand Vizier, for all his apparent agreement with the Emperor, had disclosed other views to Lord Cowley. Plausible as the Canal scheme might be, what it really reflected, he now said, was the Viceroy's determination to achieve independence for Egypt. Hence the Porte, in the Grand Vizier's opinion, should evade a direct refusal of sanction, but should attach to it conditions so unacceptable to the Viceroy as to induce him to abandon the project. Cowley agreed, writing to Clarendon that he was 'much inclined to trust to the sagacity of the Turkish ministers to find means to baffle Said Pasha's ambition'. Publicly his general line now was no longer that the Canal was impracticable, but that it was 'all but impracticable'.

Meanwhile Paris rejoiced, with a salvo of martial salutes, at the birth of a son and heir to the Emperor, the Prince Imperial. The President of the Senate declaimed an address to the Emperor and Empress, which included the words: 'The East and the West, which have been seeking since the Crusades but are only now finding each other once more, will marry the two seas and their coasts to release a beneficent flow of ideas, of wealth and of civilization.'

Proceeding to London Lesseps found a post-war atmosphere, in which the peace celebrations had kindled a spirit friendly to France. He was presented to Queen Victoria, briefed Prince Albert at length on his plans for the Canal, and inspired the Duke of Cambridge to wish them success. He was given a dinner by the Royal Geographical Society. He was gratified to find that the geographer Wyld, proprietor of the Great Model Globe in

Leicester Square, was introducing into his exhibition a large-scale relief map of the projected Suez Canal, to replace that of Sebastopol, with demonstrations to a large public, three times a day, of its advantages over the route round the Cape.

Only Lord Palmerston still remained blandly impervious to any new atmosphere. The 'man of 1840' proved as defiant as ever in his prejudices against France in relation to Egypt. He persisted, according to Lesseps, 'in maintaining that the execution of the canal is physically impossible and that he knew more about this subject than all the engineers in Europe whose opinion will not shake his own'. The Prime Minister 'advanced the most contradictory, the most incoherent and I even dare to say the most insane views concerning the Suez Canal that one can imagine'. He saw it as a step in the Machiavellian policy which France had long been pursuing against Britain in Egypt. He delivered a long and confused tirade on the harm that it threatened to Egypt and Turkey. 'While listening to him,' concluded Lesseps, 'I asked myself whether I had before me a maniac or a statesman. Not one of his arguments could be sustained in a serious discussion.'

In any event, following these various diplomatic manœuvres in Paris and London, it was evident that the centre of gravity would now shift back to Constantinople. Politically it was on the Porte that Lesseps must again train his batteries. Technically the scheme for the Canal was now based on firm ground. Financially a sound structure had still to be built for it. Much political ground had to be covered before this could be done. But at least he was starting to lay its foundations, in Paris, in London, elsewhere in Europe—and in Egypt itself.

CHAPTER VII

Towards the *Fait Accompli*

Early in January 1856, following the preliminary report of the International Scientific Commission, Lesseps obtained from the Viceroy a new Concession, on the lines of the old but with certain amplifications and changes of detail. No mention was now made of an indirect route, only of the direct route for the maritime canal. The Company was now entitled to fix the Canal dues up to a certain maximum, independently of the Egyptian Government. Provision was made for the renewal of the Concession, after ninety-nine years, subject to an increased percentage to the Egyptian Government. Lesseps himself was to remain president of the Company for ten years from the opening of the Canal. It was stipulated that four-fifths of the workers employed in the Canal should be Egyptians. No mention was now made of fortifications by the Egyptian Government.

The statutes of the Company, published at the same time, established its headquarters in Alexandria and its legal domicile in Paris. Its capital was fixed at 200 million francs—£8 million—divided into 400,000 shares at 500 francs each. Each holder of twenty-five shares had the right to vote at the Annual General Assembly, with a maximum of ten votes per shareholder. The Assembly would appoint a Board of Directors of thirty-two members, representing the nations chiefly involved in the enterprise. The Board was given wide powers, largely exercised in practice by an Executive Committee composed of its President and four of its members.

Finally the report of the Scientific Commission was followed by the publication in Alexandria of a preliminary list of Founder Shareholders, as approved by the Viceroy. The ultimate issue for this purpose was a thousand shares, divided among 170 persons. Lesseps himself, together with Ruyssenaers, Linant and Mougel, were each allotted twenty shares—a modest investment of 10,000 francs. Other investors, composed largely of his own friends and friends of the Viceroy, had holdings of 2,500 or 5,000 francs each. There was little difficulty in raising this initial capital 'in a country', as Bruce expressed it, 'where Government business is eagerly sought after, and where the calculation is that zeal shown in favour of a scheme patronised by the Viceroy will be remembered when commissions are to be bestowed'.

When the time came for raising the bulk of his capital, it had always been Lesseps's laudable if naïf intention to distribute the Canal shares among the largest possible number of small investors in all countries. He had even less love for 'the vultures and lynxes of finance', as he called them, than for politicians and their devious ways. Friends had warned him that he could not dispense with the big financiers. Nothing, he agreed in reply, would be easier than to do a deal with them to his own profit and theirs. But at whose expense? That of the ordinary investing public; that of the credit of the Company itself, whose name would thus become soiled by the taint of corruption.

Thus, whether in fact from his own choice or theirs, Lesseps's few contacts with the big banking-houses in Paris came to nothing. Advised to seek the support of his old friend Fould, whom he had consulted with regard to his earliest project for a Canal, he replied: 'One cannot come to an understanding with those people. Bankers want to lay down the law to me, and I won't have it. I shall carry out the business all by myself, and I shall approach the public direct.'

He explained his scheme, and his intention to raise capital amounting to 200 million francs, to Baron James de Rothschild, to whom he had done a service as Minister in Madrid. Rothschild reacted enthusiastically to the project, and offered him the services of his offices in Paris and abroad; also of his publicity organisation, when Lesseps should have completed his preparatory negotiations and studies and the time came to launch the subscription. Meanwhile he gave him introductions in the City of London, where he made contact with the Barings and other

Ferdinand de Lesseps bestrides his Canal

Thomas Waghorn, originator of the Overland Route

banking circles. Lesseps was delighted with Rothschild's offer, thanking him and, as a simpleton in matters of finance, asking him only as an afterthought what these services would cost him.

The Baron replied with a laugh: '*Mon Dieu!* It is easy to see that you are not a business man. It will be the usual five per cent.'

'Five per cent!' Lesseps exclaimed. 'On 200 millions? But that's ten million francs! Very well, I'll find premises somewhere for 1200 francs per month and manage the affair just as well on my own.'

Lesseps was eventually to establish such an office in the Place Vendôme in Paris. But three more weary years of frustration and conflict were to pass before he could do so, before his Company could be floated as a registered business concern, ready to exploit with its own capital the Concession entrusted to him.

*   *   *

Meanwhile the political situation confronting it had crystallized all too clearly. Lesseps found himself in an impasse. British official opposition was implacable. So, as long as it persisted, was that of the Porte. Juridically, Lesseps held the view that the consent of the Sultan was not essential to the Canal scheme. Politically however it was still, if only in the view of the Viceroy, desirable.

Through these years the idea grew in Lesseps's mind that he might have to force the issue, proceeding, regardless of the Porte, to a *fait accompli,* with the flotation of the Company and the start of constructional work. But the process should not be hurried. Too much was at stake to permit of too many chances being taken. With an enterprise so vast in its implications and world-wide in its scope as the junction of the seas, Lesseps was reluctant, until every effort had failed, to prejudice the issue through a step so drastic and of such unpredictable consequences. Thus, drawing on his deep resources of patience, he moved slowly and industriously, step by step, in a continuous endeavour to mobilise his supporters, to overcome his opponents, and thus to open the way for the start of his enterprise.

His tactics were first to enlist official support among the other powers of Europe, and so isolate Britain; secondly, to stimulate unofficial opposition in Britain itself to the British Government's policy. Finally, when the time was ripe but not before, he would seek to enlist the Emperor's official support and that of his Government, which was still not forthcoming. In pursuit of these ends Lesseps brought into play the full armoury of his propaganda machine, writing countless letters, drafting countless memoranda, making countless speeches; and he travelled without respite, backwards and forwards between Cairo, Paris, London, Constantinople and the countries of Europe.

Here his first hope lay with Austria, a power whose Mediterranean trade gave her a positive interest in the route to the East through Suez. Metternich, 'the illustrious doyen of European diplomacy' as Lesseps described him, had been an active protagonist of the Canal since the time of Mohammed Ali; he had written to the present Viceroy to wish it success; and it was he, with the Austrian Finance Minister (a founder shareholder) who had encouraged Lesseps, through Count Buol, to make a move, at the Peace Congress, for its neutralization.

On leaving Paris Lesseps visited him in Vienna. In the course of a long talk Metternich, for all his eighty-three years, showed a clear grasp of the background of the scheme since the time of Napoleon I, and talked lucidly on the subject, assessing its various factors in relation to the present interests of the powers involved. He held that the Viceroy had every right to order the construction of the Canal on his own authority, but that, in view of these interests, he had acted wisely in requesting the Sultan's sanction. This, in the case of an enterprise so beneficial to the Ottoman Empire, ought not to be in doubt now that scientists had pronounced in its favour and sufficient capital was likely to be available for its execution.

Metternich made an important distinction between the two aspects of the problem, which in his view should be separately handled. The construction of the Canal was a domestic matter, involving only the Porte. Its neutrality, on the other hand, was an international matter, involving the powers in concert. To discuss the settlement of this in perpetuity, under the terms of the Concession, the Sultan should invite them to send plenipotentiaries to a conference in Constantinople. Such a course would at once safeguard the sovereignty of the Porte, ensure to it, following the Peace Congress, the position of influence which it merited in face of the political and commercial interests of its allies in Europe, and provide a fresh guarantee of its future integrity and independence. And this would deprive Britain of all motive and pretext for interfering in the internal affairs of the Porte by blocking the construction of the Suez Canal.

Such, as Metternich described it, was his 'political testament', for transmission through Lesseps for the approval—which was not in doubt—of the Viceroy of Egypt. Unfortunately this enlightened and statesmanlike plan presupposed the acquiescence of the Porte, which in turn presupposed that of Britain. At present the evasions and devious obstructions of the Porte created a barrier as effective as the overt opposition of Britain. This did not deter Lesseps from pressing Metternich's testament wherever he went. But it received only qualified support from Austria's missions abroad, whose main concern was to conciliate Britain; and it was to fall into abeyance with the invasion of Austrian territory by France, in Italy, in 1859.

In these peripatetic years Lesseps made three visits to Constantinople. But his batteries had little effect on the Porte. Lord Stratford, with lofty irony, would announce to London the reappearance of 'the celebrated Mr. Lesseps'. ('He has done me the honour to call, but I was not at home') He was coming 'to do his worst'. Thanks, however, to the return to power of Reshid, his old ally, as Grand Vizier, Stratford did not 'entertain any serious apprehension of his succeeding as far as obtaining the Sultan's consent, but I doubt whether a decided refusal will be given'.

Lesseps's line was to make it appear that rejection would be 'unpalatable to the Tuileries'. But in fact he had little or no encouragement from his own Government. Sabatier, in Alexandria, had been rebuked by the Quai d'Orsay in response to a complaint from the Foreign Office, for exploiting the Emperor's name with the Viceroy to further the Canal. Lesseps had been similarly warned by Walewski to expect no assistance as long as the British Government opposed the scheme, and his visits to Constantinople were a source of some embarrassment to Thouvenel, the French Ambassador.

Hope was to spring when Stratford, at the end of 1857, retired from his post after a decade of continuous service as British Ambassador. Arriving at this propitious moment, Lesseps proposed a plan by which the Porte should invite the powers to agree among themselves before committing itself to a decision—thus shifting the onus on to Europe and placing Britain in an invidious minority of one. After a 'reconciliation dinner' with Reshid and Thouvenel, he set to work to convert the Turkish ministers 'between the coffee and the pipe' to his new idea. But, despite a long and encouraging talk with Reshid at his villa on the Bosporus, followed by a persuasive memorandum, quoting Metternich's formula, nothing materialized. The British line was to frighten the Ministers with the idea that the Canal was a French military instrument—'a wide, deep and defensive ditch'—designed to separate Egypt from Turkey.

The British Government went so far, on New Year's Day 1858, as to deliver a threat to the Sultan that his consent to it would reverse Britain's traditional pro-Turkish policy. 'His Majesty must not expect that the maintenance of the integrity of the Ottoman Empire could thereafter be a principle to guide the policy of the Great Powers of Europe, because the Sultan himself would have been a party to the setting aside of that principle.' In response to this warning, the Turkish Foreign Secretary gave a verbal assurance that his Government would not consent to the Canal without the British Government's sanction.

*   *   *

During these years, between journeys, Lesseps returned whenever possible to Egypt, to supervise the preparations of his engineers and above all to fortify the Viceroy's morale. Mohammed Said was often discouraged—by the endless delays; by agents of the Porte, working to discredit him with his subjects and in particular his Army; by the pressures of his British Consular adversary and the doubts which he insidiously sowed in his mind. Bruce, on instructions, was pressing on him the dangers of French control over his Government, which would weigh more heavily upon him and prove more galling than his present connection with the Porte. He carried *lèse majesté* to the point of suggesting that, if the Canal were entrusted to a foreign company, the Viceroy, with a large province in the possession of foreigners, would become no more than its *Vekil,* a remark at which Said 'made a movement on the divan as though he had been stung.'

Thinking wishfully, Bruce reported to London a few months later that if the Sultan refused his consent, the scheme would be dropped for the present and the Viceroy's attention would be switched to other public works: 'He has no doubt cooled in his ardour on the subject, and will probably be secretly not dissatisfied to be relieved from this incubus.' It was the general view in Alexandria that the Canal scheme was petering out. This hardly went so far as the view of Lord Clarendon however, as expressed to Stratford, that, once Lesseps had completed his fortified ditch, 'at the expense of the dupes who might be persuaded to put their money into the speculation', he would declare the completion of the work impossible and proceed to wind up, 'having attained the political object in view, or at least the essential part of it'.

Beset by such worries, the Viceroy welcomed Lesseps with evident relief, exclaiming, 'Your arrival reconciles me with humanity.' It was their habit to make a joke of their trials with the British. Once, when Lesseps saw him after a long interval, the Viceroy, whose clothes hung loosely upon him, remarked: 'See how thin the English have made me!' Once Lesseps made him a present of a cane, which Said always kept beside him, together with another, presented to him by a British admiral. He made a pact with Lesseps that at their interviews, when the company in attendance made it inopportune to talk of the Canal, he would carry the English cane. But if he was carrying Lesseps's cane, the coast was clear for Lesseps to talk of it as freely as he wished.

Now, to get away from it all, the Viceroy bore him up the Nile on a three-months' tour of the Sudan. The voyage had an ominous start when the mosquito-net in Lesseps's cabin caught fire, and he was painfully burned. But he turned the omen into a good one, suggesting to Said that their debts to the evil fates were now discharged. Their old intimacy was resumed, Said confiding in Lesseps, Lesseps giving him advice on his vast administrative problems. This drove the Viceroy one evening to an outburst of fury and self-reproach at some error, from which he recovered next morning with the reassurance: 'On your return, you will see, you will be pleased with me.' Altogether, the Sudan voyage, as Lesseps wrote to his mother-in-law, would 'serve the Canal'.

*   *   *

On their return it was time for Lesseps to proceed once more to London, on a tour which, so he assured the Viceroy, should prove decisive. Stumping the British Isles for seven weeks in May and June 1857, he visited sixteen cities and addressed twenty meetings, organized by local Chambers of Commerce or by groups of ship-owners, merchants, bankers, and industrialists. Helped, as translator, by his British agent, Daniel Lange, he spoke with persuasive warmth and good humour, illustrating his arguments from an array of maps, plans, and statistics, and debating with forceful conviction. He asked for no funds, only for approval in principle; and these hard-headed men were impressed, supplying him with a large dossier of agreed and signed resolutions in favour of the Suez Canal.

Lesseps was flushed with the success of his campaign. To the support of the engineers of Europe, he had now added that of the commercial community of Britain. With this he hoped to establish the fact 'that the enterprise of cutting the Isthmus of Suez will be profitable to English interests, and that no Government has the right to oppose it'. He had yet to realize that Britain was two nations, industrial and political, and that the one still had, for his purposes, all too little influence over the other.

Returning to London he was again invited to an evening party at Lady Palmerston's. Palmerston accosted him with the remark: 'Well, you have come to make war on us in our own country. You have stirred up England, Ireland and Scotland to agitate for the Suez Canal.'

'Certainly, my lord,' Lesseps replied, 'I have profited by the liberty which I admire in your country, since it allows me to speak freely everywhere, in public, on subjects which don't at all please the Government.'

Lord Palmerston retorted: 'You know I am quite frankly opposed to your scheme.'

'I think', said Lesseps, 'that the public opinion which I have just come to recognize will overcome individual resistance, as happily occurs in this country. I must add that here, more than anywhere, opposition is one of the greatest elements of success.'

That morning, he continued, he had seen a striking example of this. Wishing to publish, for wide distribution, copies of the reports of his meetings, he had asked the editor of an important newspaper for an estimate as to how much this might cost. When the editor gave him the figure, Lesseps observed that an especially large sum had been allowed for an article to oppose the publication.

'Very well,' replied Palmerston, 'you don't grudge me my opposition. I am glad it doesn't upset our relations.'

'I grudge it so little', Lesseps jested, 'that if I had a hundred thousand francs to give you for every speech you make against the Canal in the House of Commons, and if you were the kind of man to accept, I should hasten to offer them to you. For it is your opposition that will bring in the capital which we need for our enterprise.'

None the less, the success of Lesseps's tour had made its impact on Palmerston. The moment had come for a public pronouncement, in response to his jesting challenge. A few weeks later, on July 7, 1857, the question of the Canal was raised for the first time in the House of Commons. The Prime Minister was asked a question about the scheme by Mr. F. H. F. Berkeley, the Member for the City of Bristol, whose merchants had passed a strong resolution in its favour. Would the Government use its influence with the Sultan to obtain his sanction for the Canal's construction, requested by the Viceroy, and approved by 'the principal cities, ports and commercial towns of the United Kingdom?'

Lord Palmerston replied grimly in the negative, pointing out that 'for the last fifteen years Her Majesty's Government have used all the influence they possess at Constantinople and in Egypt to prevent that scheme from being carried into execution'. It ranked 'among the many bubble schemes that from time to time have been palmed upon gullible capitalists', and he strongly advised the Honourable Member and his constituents to have nothing to do with it. After repeating the fallacy that it was 'physically impracticable'—except, he conceded, at too great a cost—he added that it was in every way 'adverse and hostile to British interests'. Liable to facilitate the separation of Egypt from Turkey, it was opposed to the country's settled policy, which had been consecrated by the late war and the subsequent Treaty of Paris.

Some sections of the Press deplored Palmerston's personal aspersions on Lesseps's character as the promoter of a 'bubble scheme'—*The Spectator*[1] remarking that Parliamentary privilege protected him against an action for libel. *The Bristol Advertiser*[2] condemned his remarks as 'an uncalled-for and unmanly impertinence'. In general Palmerston's answer was full 'either of that irredeemable confusion which is that fruit of ignorance, or of that wanton misrepresentation which some will praise as a triumph of diplomacy'. His 'off-hand, flippant, and inconsistent notice of the scheme was no proof of its inexpediency on patriotic, its inutility on commercial, or its impracticability on scientific grounds'.

The *Daily News*[3] refuted Palmerston's views in two long articles; the *Commercial Record*[4] expressed astonishment that political power should so dim the intelligence of an otherwise perspicacious man; while the *Mercantile Journal*[5] concluded that

[1] July 11, 1857.   [2] July 11, 1857.   [3] July 9 and 10, 1857.   [4] July 31, 1857.
[5] August 8, 1857.

Britain's bankers were, in their ideas, a generation ahead of her rulers.

Some days later Palmerston was pressed once more in the House, this time by Mr. Darby Griffith, who enquired whether 'it be conducive to the honour or the interests of this country that we should manifest and avow the existence of a jealous hostility on our part to the project of a Ship Canal through the Isthmus'. Whatever the engineering problems, he could not understand why this should be contrary to the country's interests. Palmerston, after repeating his previous political arguments, contended that in connection with our Indian possessions the Canal would benefit other naval powers at our expense, giving them 'a very important start as compared with ourselves with regard to any operation that might be undertaken in the Indian Seas'. He defiantly reiterated that the project, even if practicable, could not pay, and asserted again that it was 'one of the bubble schemes which are often set on foot to induce English capitalists to embark their money on enterprises which, in the end, will only leave them poorer, whomever else they make richer'.

On the engineering aspects of the problem, Palmerston received support from Robert Stephenson, who recounted at length his own investigations in the Isthmus, maintained that the scheme was, if only commercially, impracticable, and concluded that his own railway, now nearly completed, would 'as far as concerned India and postal arrangements, be more expeditious, more certain, and more economical' than 'this new Bosporus between the Red Sea and the Mediterranean'.

Stephenson's speech aroused the Gallic wrath of Lesseps, back in Paris, when he read a report of it in *The Times,* implying as it did an endorsement of Palmerston's aspersion on his business integrity. He immediately hastened across the Channel, and demanded a written explanation, either by Stephenson himself or through two chosen friends. This amounted to a challenge to a duel, which he was debarred from issuing to the Prime Minister in person. But Stephenson returned the soft answer, disclaiming any intention to attack Lesseps personally, and explaining that his support of Palmerston covered only his assertion that the

Canal scheme, if ever completed, would not be 'commercially advantageous'.[1] So the matter was dropped.

*　　　*　　　*

During that summer of 1857, Palmerston's political arguments were dramatically put to the test, and the Suez route placed in a new perspective, by the outbreak of the Indian Mutiny and the urgent need for the rapid despatch of troops to India. At first they were sent by the long route round the Cape, largely in transports propelled by sail. This situation was exploited by the mutineers, proclaiming that the Sultan of Turkey had ordered the closure by his Viceroy of the route to India across Egypt, that there was thus no need to fear the arrival of British troops, and that Canning, the British Viceroy, was plunged in despair at this news. In fact Said, prompted by Lesseps, offered full facilities for the transit of troops across Egypt. The railway from Alexandria was now far enough advanced to carry them for much of the way, and it was for just such an emergency that the Overland Route and the railway itself had been designed.

Parliament put pressure on Palmerston to adopt the route, at first without success. He argued that the disembarkation and re-embarkation and the land journey across Egypt would expose the troops to undue fatigue. But in the autumn he relented, and the P. & O. started to send steam transports to Alexandria and from Suez to Calcutta, carrying preserved meats to provision them on their journey across the desert—though at first the route was limited to reinforcements of engineers, artillerymen, and medical services. This step was exploited by the Opposition Press as a weapon with which to condemn the antagonism of Lord Palmerston and Lord Stratford de Redcliffe, to the Suez Canal.

Some weeks later, at a banquet in Lesseps's honour in Vienna, the Austrian Finance Minister, Baron de Bruck, proposed a toast

---

[1] Four years later he similarly challenged Lord Carnarvon, through two 'seconds', a General and an Admiral, for declaring in the House of Lords (May 6, 1861) that the Suez Canal Company was bankrupt. But once again the affair blew over.

to the victory of British arms in India, remarking: 'If, from the start of the mutiny, the Isthmus of Suez had been pierced and the ships had been able to take the direct route, what torrents of blood could have been saved and what barbarous acts prevented! Here surely was a new motive for aiding with perseverance and zeal this finest work of civilization.'

*   *   *

Early in the following year, 1858, an Anglo-French crisis arose over a plot hatched in Britain by Orsini and a group of political refugees from France, to assassinate the Emperor Napoleon. The fact that the conspirators had been able to obtain bombs and other weapons and to lay their plans in Soho, under cover of British law, unleashed in Paris a storm of popular indignation against Britain, and Walewski was obliged to despatch a sharp note to Clarendon, with a request to curb the activities of the refugees. Palmerston introduced a Conspiracy to Murder Bill, to strengthen the law by treating such cases as felonies. But this was opposed, both in Parliament and in the country, on the grounds that it truckled unduly to pressure from France. The Government itself was divided. On a vote in the House, it was obliged to resign, and a new Cabinet was formed by Lord Derby.

Hope sprung again in the mind of Lesseps, now in Constantinople, as it had done on the recall of Lord Stratford, followed shortly afterwards by the death of his adversary Reshid Pasha, the Grand Vizier. Now Palmerston himself had fallen. Here at last should be his chance for the Suez Canal. A change of Government in Britain must surely lead to a change of opinion and policy. He was soon disillusioned. On enquiry, the Turkish Ambassador in London was informed by the new Foreign Secretary, Lord Malmesbury, that his Government intended to pursue precisely the same course as its predecessor. It became clear to Lesseps that the policy of Britain was the product not of the minds of a few powerful individuals but of the reasoned working of the national political machine as a whole—and this made it all the more redoubtable.

The mouthpiece of the Government on the subject of the Canal was now Disraeli, the new Chancellor of the Exchequer. He had a chance to express his views on April 12, 1858, when a question was asked in the House by Mr. Griffith. This time Griffith confined his enquiry to the Government's political objections to the Canal—'recondite political speculations of so finely drawn a character as to be not at all obvious to the comprehension of ordinary mortals'. To this Disraeli replied in more qualified terms than those of Palmerston. The present Government, he said, had no evidence before it to justify opposition to the Canal on political grounds. Only after seeing evidence that the Canal was practicable and commercially desirable could he express a political opinion concerning it. Meanwhile he believed this to be 'an operation which could only end in failure'.

He spoke at greater length on the subject in a full-dress debate on June 1. This time a motion was introduced by Mr. J. A. Roebuck, the Member for Sheffield, a leading Radical and an inveterate antagonist of Palmerston. What he desired was a declaration that 'in the opinion of this House, the power and influence of England ought not to be used in order to induce the Sultan to withhold his consent to the project of making a Canal across the Isthmus of Suez'. Here we were behaving, he maintained, in 'a selfish and base manner'. On general principles he insisted that 'facility of transport from one part of the earth's surface to another was for the benefit of mankind at large'. He deprecated the proposition that 'What was for the interest of mankind was not for the interest of England'. He contended that the House of Commons had no concern with the physical or commercial aspects.

No British political interest, he believed, could be injured by a safe and easy transit from Europe to India, where Britain had a greater traffic than the rest of the world put together. Our dominion in India depended on our maritime superiority. This had enabled us, at the time of the French invasion of Egypt, 'to shut up Napoleon and his army like rats in a trap'. It would enable us, if the Suez Canal were cut, to pursue a French fleet through it and catch them in the Red Sea like rats in a trap. 'The danger arising from the expectation that at some moment France and some other power might be superior in the Mediterranean was altogether illusory.' Our opposition had created 'a feeling in France that we were an insolent, an insular, a grasping and a selfish people'. Yet 'our Government called itself a friend of the French alliance. Was a mere crotchet of Lord Palmerston's, as some had called it, to stand in the way of the execution of the greatest physical work ever undertaken since man was upon the face of the earth?'

Mr. Griffith, following Mr. Roebuck, pushed his arguments home by pointing the recent moral of the Indian Mutiny: 'If at that time we could have embarked the troops in our large steamers, and conveyed them right on to India by the Red Sea, without any break in the transit at the Isthmus of Suez, how much of anxiety would the country have been spared, while the mutiny might have been divested of many of its horrors.' Lord Palmerston had dwelt on the difficulties of this route, but they had vanished in practice, and now we were using it regularly, to send troops to India.

Mr. Stephenson weighed in, as before, on the Government side, with his technical 'expertise'. Lord Palmerston, speaking as a private member, reiterated his well-worn arguments against this 'great military canal' which was also a great commercial bubble, raking up another from the past to the effect that the Red Sea winds would make the passage of sailing vessels [*sic*] through the Canal 'inconceivably slow'. After a number of other speeches from both sides of the House, the debate ended in a confrontation between Gladstone and Disraeli, winding up for the Government.

Gladstone rose, not to request Government approval for the scheme but to protest against the use of the political influence of this country against it. The House was being asked to 'put an end to the vicious system . . . of arbitrary and gratuitous interference for the purpose of preventing the execution of this Canal, on grounds which are either null or valueless, but which are in reality much worse because they go to place England at issue with the world, and to commit us to a contest in which we must necessarily fail'. Originally the official objections to the French

Canal arose from the fear of its competition with the English railway. In fact the same arguments applied to both: 'The possession of the railway would be an advantage to hostile powers in sending troops to India almost as great as the Canal.'

Gladstone based his case on the fact that this was a scheme 'beneficial to mankind'. He was concerned neither with the technical nor with the commercial but only with the political objections—the supposed dangers of dismemberment both of the Ottoman Empire by the separation of Egypt from Turkey, and of the British Empire by the separation of Britain from India. Dealing with the first, he said: 'Egypt is subordinate to Turkey, not on account of the strength of the Sultan but on account of the interests of Europe and of the Europeans'; and British opposition to the Canal on these grounds was 'utterly unsatisfactory to the whole of Europe'. Dealing with the second, he said: 'I am unwilling to set up the Indian Empire of Great Britain in opposition to the general interests of mankind, or to the general sentiment of Europe.' Such opposition was 'of a kind calculated to create in Europe a sentiment of irritation, of jealousy, and even of hostility to the existence of British power in India'. Gladstone concluded:

> The policy which has been pursued is not only a false policy, but is so diametrically opposed to the first principles of prudence, and I will even say to the comity and courtesy of friendly nations; it has such a tendency to isolate us on these questions from the rest of civilized mankind—and this fact will come to be increasingly felt from year to year by the British people—that you will not be able permanently to maintain it. If then you are to recede from this policy, the sooner it is done, the better.

Disraeli, rejecting the motion on behalf of the Government, took a warier and less positive line than Lord Palmerston had done. There was no evidence, he declared, that either this or the previous Government had used any improper constraint on the Sultan, or had exercised what Gladstone described as an 'improper, undue and illegitimate opposition' to the Canal scheme.

No opposition could be such if it spoke for the interests of Britain. Lord Palmerston 'can only have enforced on the Sultan the pursuance of a policy which the Government of the Sultan had already professed', and it was a subject on which the Porte was known to have decided opinions of its own.

Nor was there any evidence 'that this project has the approval of all other nations'. On the contrary, France had shown 'a very candid sense of the difficulties of the whole project' and had shown no wish to interfere in it. His conclusions were guarded:

> There are grave considerations connected with this question both as regards Turkey and as regards England. I do not say that their gravity is such that they may not yield to other considerations, if in time those other considerations are deemed more powerful; but, as at present advised, I think it would be a most rash and precipitate step to pledge this House to change the course of policy which it has long pursued, and which has been sanctioned by high authority, and to ask us to place ourselves in a position which will prevent us hereafter from adopting that line which we may deem wisest and most prudent.

Lord John Russell, supporting Gladstone, gave the lie to Disraeli, declaring it to be 'an admitted proposition' that 'for many years the power and influence of this country have been used to induce the Sultan to withhold his assent' to the Canal. On the commercial issue, he spoke for free trade, believing it beneficial to England 'to subject ourselves to any competition by which the commerce of the world can be increased'. On the military issue, he saw no danger as long as England had command of the seas.

Mr. Roebuck, winding up the debate, ridiculed the suggestion that the Sultan had not been coerced: 'The power of England, exercised through one of the most imperious Ambassadors, could not be properly designated otherwise than as coercion, and all he asked was that the Sultan should be left to himself.' His motion was rejected, by 290 votes to 62. The 'liberal and commercial spirit of the age', as a speaker had called it, was not yet a match for its more conservative political elements. They preferred, in the light of the policies of an earlier age, to play for safety and time over a project so radical scientifically, so far beyond the present range of their vision, and so coloured politically by their ingrained mistrust of the French.

Lesseps was not as discouraged as he might well have been by the result of the debate. At least the project of the Canal had made an impact on public opinion. The quality and persistence of its supporters in the House had achieved for it a moral success. In his optimistic assessment there was a general opinion that 'the progress of the Company cannot be stopped, and that the opposition is untenable'.

But this was a long-term view. The door might not have been irrevocably closed, but it was only just ajar, and was unlikely to open in the predictable future. Meanwhile Disraeli's speech clinched Lesseps's resolution to proceed at last towards the *fait accompli*. The Porte, as he had learned in the course of his recent prolonged visit, was less disposed than ever to grant its sanction. The Austrians, always wavering towards the line of least resistance, had now started to echo the English about 'bubble schemes'. The official attitude of France, whatever the Emperor's personal views, remained as negative as ever.

Thus Lesseps, despite the disagreement of his more cautious associates, decided to go ahead with the organization of his Company on the basis of the concessions which the Viceroy had granted him. He proceeded to Egypt, where he turned all his powers of persuasion on the Viceroy. This course, he argued, would ensure overt support for the Canal from the majority of the European powers, who already favoured it in principle. The Viceroy appeared to agree, though he was later to imply that he had not done so.

Lesseps made a last propaganda tour throughout Europe, canvassing supporters, enlisting agents, and inviting subscriptions, in such centres as Venice, Barcelona, Odessa, Trieste and Marseilles, and despatching persuasive circulars to other influential cities. The policy of the *fait accompli* was launched. As Lesseps expressed it to Ruyssenaers: 'I have just raised the curtain of the last act.'

Mohammed Said Pasha, Viceroy of Egypt

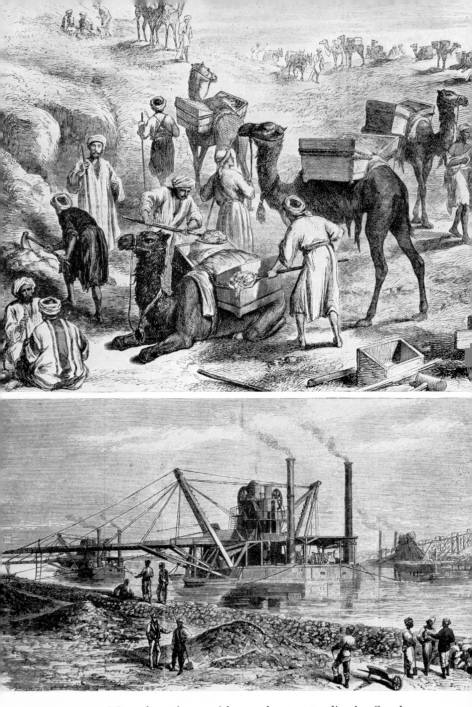

Manual workers, with camels, start to dig the Canal

Mechanical dredgers and elevators complete it

CHAPTER VIII

Formation of the Company

On November 5, 1858, the subscription list for the shares of the Suez Canal Company was opened to the public. Each share carried a guaranteed interest of 5 per cent throughout the period of the Canal's construction. A little over half the total number of 400,000 shares were taken up in France by some 21,000 shareholders. The great majority of them were small investors—*rentiers* and merchants, engineers, lawyers, teachers and doctors, officers, civil servants and artisans. Analysing the list, Lesseps saw, as he had hoped to see, the stirring of popular instincts, middle-class sentiments, military patriotism and the intelligence of the liberal professions.

In London, *The Globe* saw it differently: 'The principal subscribers are hotel waiters, who have been deceived by the papers they have read, and petty grocery employees who have been beguiled by puffs. The priesthood has been victimized and 3,000 day labourers have been induced to pool their savings to buy these shares. . . . The whole thing is a flagrant robbery gotten up to despoil the simple people who have allowed themselves to become dupes.' Some of them were indeed simple enough. Lesseps told of one who came to his office, wishing to subscribe to this 'railway in the island of Sweden'. When told that it was a canal in the Isthmus of Suez, he replied: 'That's all the same to me, as long as it's against the English.'

Less reassuring than the French subscription was the lack of international response to the issue, on which Lesseps had counted. Of the remaining shares some 50 per cent had been reserved for subscription through his agents and bankers in Britain, Austria, Russia and America. But none were taken up. Britain's failure to invest was only to be expected in view of the general political attitude towards the Canal scheme. Moreover the particular moment, as Lesseps had been warned, was unpropitious. For a wave of anti-French mistrust had succeeded the Orsini incident. Austria's defection was a more serious disappointment. But here too clouds were gathering, which were soon to break with Napoleon's war in Italy against the Austrian Empire. In this atmosphere of political tension Austria was strengthening her relations with Britain, and her Government was not at present prepared to underwrite any issue of Canal shares by the Austrian bankers. Hence, after a series of small subscriptions had been taken up in a dozen other countries, including Spain and Holland, the Company was still left with 176,642 unsubscribed shares on its hands.

This put Lesseps in an awkward position. For the Company could not be legally registered until the whole of its authorized capital had been subscribed. Had he been prepared, in the first place, to rely on big French bankers and not only on small shareholders for his capital, this problem might not now have arisen, for their participation, supplementing and supporting the public investment, could have covered any such difference. Lesseps had avoided this course because he felt the need for independence and personal freedom of action; because he was a sincere believer in the international principle; because he lacked all experience of the hard realities of high finance. A man totally uninterested in money as such, he preferred unorthodox financial methods—which, to some, seemed unscrupulous.

The Viceroy, as ignorant in respect of finance and as bent on his own independence, had generally favoured his view. Hence it was to the Viceroy that Lesseps now had to turn for his unsubscribed capital. Already Said had promised to take up 64,000 shares. In conversation with Lesseps, before the formation of the Company, he had also agreed to cover any foreign subscriptions which were not taken up. Unfortunately he had given no written undertaking to do so. Nevertheless, confident that this was his genuine intention, Lesseps now took up the remaining shares on the Viceroy's behalf, and so secured the registration of the Company.

Mohammed Said, on the other hand, was taken aback at the magnitude of the sum involved by his impulsive promise. It amounted to 44 per cent of the Company's total registered capital, an ultimate investment amounting to some 85 million francs (£3,500,000) which was greatly in excess of his anticipations. This was to cause him doubts and misgivings for some years to come, with consequent worries to Lesseps. These were only partially relieved by agreements to delay the necessary payments by spreading them over a number of years. Said, though he might grumble and hesitate and object, at no time repudiated his commitment. But in fact it was not until after his death that it was officially accepted and signed.

These arrangements laid the Company open, from the start, to uneasy criticism in financial circles, providing fuel to the Canal's opponents by enabling them to cast doubts on the financial reliability of Lesseps and his associates. He had in fact committed an irregularity in applying for the registration before the authorized capital was fully subscribed—though this, through his resort to the Viceroy, was soon remedied. Again, he was criticized for opening the subscription before naming the members of his Administrative Council. To this he replied that it must represent the shareholding nations whose response to the flotation was greatest. Though there was no such response, and the Company was now virtually French, he still gave the Council, in his capacity as President, as international a character as possible.

In response to a further criticism that among its French members were too many of his own friends and relations, he argued robustly that he did business with those who were with him and not with those who were against him. As patron of the enterprise he secured Prince Jérôme Bonaparte, the Emperor's uncle—to be succeeded on his death by his son, Prince Napoleon. The technical direction of the Canal itself, for which responsibility had hitherto been shared between Linant and Mougel, was now entrusted to Mougel alone, as Director-General of Works—the sounder engineer of the two for the executive task which now immediately confronted the Company. For, as soon as it was formed, Lesseps decided that the works were to start, if only on the site of the Sweet Water Canal, which was a necessary feeder to the Isthmus.

This at once caused the Viceroy alarm. He had accepted the formation of the Company, though disconcerted by its unforeseen demands on his Exchequer and by reports of its irregular constitution. It could be argued, as Lesseps had done, that this was within his rights as a vassal of the Sultan. But the start of the works was another matter. Though this was the next logical step towards the *fait accompli,* Said shrank from taking it without the Porte's approval. He assured Green, the Acting British Consul-General, 'that not a sod shall be turned either with relation to the fresh-water canal or the ship canal until the Sultan's sanction shall have been obtained', and reminded him that this had been a condition attached to the original Concession.

Mohammed Said was in an awkward predicament. In his mind the Canal scheme had started as a private dream for the glory of his name and his country. But as it turned to reality it had become a contentious international issue, compromising his relations with the powers, whom he sought to please and impress, and above all his relations with the Porte which, for a very good reason, he needed their aid to conciliate. Said had no very high opinion of the Turks—'those idiots', as he described them to Green, 'sitting at Constantinople with their mouths and hands open and their eyes turned up to Heaven, waiting for a miracle to save them from ruin'. Did they not hold a contemptible position in the eyes of all nations, even of the Sardinians, the Spaniards, the Maltese? Nor had he any especial personal respect for the Sultan.

But he sought from him one major concession, which his predecessor Abbas had sought in vain. This was the consolidation of his dynasty by the right of direct succession from father to son, in place of the indirect succession through the oldest living collateral relative, which meant that the Viceroyalty would pass at Said's death to his nephew, and near-contemporary, Ismail. 'When I look at that cousin [*sic*] of mine', he exclaimed to Green, 'I ask myself whether I am not a fool to labour for such a creature as that, with a large shawl round his throat, afraid of going over a bridge.'

Much as Said desired the Canal, he had always tended to see it not only as an end in itself but as a means to this other end. By uniting the two seas he could raise his prestige in Europe and so gain the support of the powers for a change, through the Porte, in his law of succession. Such was his ultimate ambition; and now ironically the Canal, far from favouring it, was rebounding against it. Hence his present predicament. In planning the Canal he had enlisted the support of France. This had antagonized Britain, and Britain had turned the Porte against him. And now France was failing to give him the support he required against both.

Said's vanity had at first been flattered by British interest in the Canal, as purveyed to him through Lesseps's well-edited accounts of his various meetings, and he saw himself cutting a figure in the outside world. Then he had grown despondent at the reports of the House of Commons debates and in particular of Palmerston's speeches. And now, following the formation of the Company, British opposition had redoubled in force and assertiveness.

The Times accused Said of plotting with France to break with Turkey, while other sections of the Press cast doubts on his sanity. The Foreign Office instructed Green to try to persuade him to abandon all idea of making the Canal. In Paris Cowley confirmed to Walewski that the British Government adhered to its policy, and warned him of disagreeable consequences if Lesseps started the works without the Sultan's sanction; and Walewski instructed Sabatier, his Consul-General in Egypt, to adhere to his own previous instructions. As before, he should remain neutral in his attitude towards the activities of Lesseps, who was none the less always to be sure of a favourable welcome at the French Consulate-General.

In London Malmesbury, the Foreign Secretary, expressed his conviction 'that Mr. Lesseps is seeking to induce French subjects to embark their capital in a vain speculation, which will entail ruin on all who may become shareholders whatever may be the amount of the pecuniary advantage which Mr. Lesseps may succeed in realizing for himself.' It was widely believed, in the light of such prevalent suspicions, that Lesseps intended to go through the motions of starting the work, and then on its prohibition to claim from the Viceroy a handsome indemnity and wind up his Company.

Alexandria buzzed with rumour and anti-Canal propaganda, and Green reported back to London his own conviction that 'the projectors of this scheme are in the position of drowning men; their situation is a desperate one; they will catch at anything to save themselves, and it is prudent to keep out of their reach'. Retailing as news a piece of local gossip, he asserted that Lesseps had written a letter to the Viceroy, 'couched in such objectionable and even threatening terms, as to have given His Highness great offence, and I have no doubt that this letter is the first step towards the *"Querelle d'Allemand"* which will wind up the proceedings'.

In fact there was no basic quarrel. Said remained in principle, if not always in practice, loyal to Lesseps and to the Canal. On his arrival in March 1859, with a small group of Commissioners from the Company's newly-formed Administrative Council, the Viceroy received him with his normal politeness. But thereafter he gave him a wide berth. Evasive by nature, it was Mohammed Said's habit, at moments of controversy, to fight shy of meetings and discussions, and withdraw from the scene. 'The best thing I can do at the moment', he said to a friend, 'is to keep aloof and to wait. . . . Lesseps has a strong back, and he will know how to cope with the situation.' To Green he said: 'I am ready to discuss the practicability, the utility and the cost of the Canal, but the political objections which have arisen to it belong to another sphere. The Governments of France and England must arrange that among themselves, or at Constantinople.' Meanwhile, until the Sultan's firman was received, he would not authorize any start on the Canal's construction. This was now his considered policy, communicated alike to the French Consul-General and to Lesseps himself.

Lesseps, realizing that he must go ahead on his own, left for the Isthmus, ostensibly to prepare his final plans but in fact to make a start, if he could, on constructional work, whatever the obstacles placed in his way. For he felt reasonably confident that, in the ultimate resort, Said would not let the Canal down, and in the meantime, whatever his official attitude, would be inclined unofficially to condone his own activities.

Before leaving Paris he had signed an agreement with a French contractor, Hardon, who had now arrived in Egypt and gave encouraging reports of the stone available for quarrying in the Attaka, the mountain ridge above Suez. Lesseps himself, with a contractor, engineers, and the Company's Commissioners, set off eastwards to follow the course of the Wadi Tumilat, the Land of Goshen, from the Nile to the dried-up Lake Timsah in the Isthmus. Here, over the past two years, the line of the alimentary Sweet Water Canal had been traced in detail. The work involved would be relatively simple, since the annual Nile flood had formed a natural bed for it; and it should lead to the rapid irrigation of large tracts of land, for the profit of the Company's share-holders.

In connection with his journey the Viceroy, equivocating between British and French, assured the British Consul-General that he had given Lesseps permission to take with him twenty labourers, not to start on the work but to assist the Canal Com-missioners in studying the ground with a view to a start in the future. Orders had been given to Governors to take action if he exceeded this number. From the start petty obstacles were placed in Lesseps's way. In Cairo—thanks, he believed, to British Con-sular pressure—the chief camel contractor refused to hire him the camels required for his caravan, and he only secured them by intimidation and a complaint to the Governor. He proceeded on his journey, in the face of obstructions ordered by the Viceroy or the local authorities or both.

As his caravan camped in the desert a Turkish officer, with a small force of *bachi-bozuks*,[1] made a descent on his mule-drivers and carried them off. Lesseps denounced him as a mere brigand, invoked the name of the Viceroy, demanded the return of the prisoners, and despatched a strong protest to the Governor of Cairo, who returned the soft answer, expressing the Viceroy's regrets and assurances of protection. But the officer continued to molest the caravan, in league with a group of local sheikhs.

[1] Irregular troops.

Lesseps, who knew his Beduin, invited the sheikhs to his tent, gave them coffee, produced a six-chambered revolver, and shot down in their presence a line of six bottles. Then he addressed them: 'My good friends, I learn that a Turkish officer, claiming to be sent by the Government, ordered you to refuse me provisions as requested this morning. I beg of you to warn this individual, who is no more than an impostor, that today we are starting off across the desert, that I have with me twenty companions, that I am not the best marksman among them, and that every black spot that we see moving in the desert we shall take for a gazelle.' Compliments were exchanged, and the desert journey completed without further incident.

None the less, interference by local officials persisted. Lesseps wrote to Said, without apparent effect, appreciating the delicacy of his position, but submitting that such actions compromised his Viceregal interests by creating the general belief that he was opposed to the Canal.

Meanwhile Lesseps and his party surveyed the irrigable lands in the Wadi, made minor changes in the tracing of the Sweet Water Canal, discovered good beds of limestone beyond Lake Timsah, inspected the quarries at Suez, and returned for a few days to Cairo and Alexandria, where they were received briefly and formally by the Viceroy. Thence they proceeded to Damietta and so across Lake Menzaleh, in a fleet of fishing boats with large lateen sails and cabins roofed with rush-matting, to the chosen site of the harbour at Pelusium, which was henceforward named, in the Viceroy's honour, Port Said. Here, from their tents by the beach, they were happy to see, moored a mile or so offshore, a brig from Marseilles, and boats unloading timber on to the sands before them—the first of many French ships to come, loaded with building materials and machinery for the Suez Canal Company.

On April 25, 1859, unfurling the Egyptian flag over the spot where the waters of the Red Sea were designed to emerge into the Mediterranean, Lesseps staged a solemn ceremony, in the presence of the Company's engineers and other employees, and for the benefit of its shareholders back in Paris. Following an appropriate presidential oration, 'on this soil, which will open up the East to the commerce and civilization of the West', the first sod of the Suez Canal was turned. One by one his companions struck at the earth with a pick-axe to open a trench where its line had been traced. Lesseps then turned to the hundred-and-fifty Egyptian workers who, unhindered by any official prohibition, stood gathered around him, and instructed each in turn to strike with his pick.

'Remember', he said, 'that you will not only be moving the earth. Your labour will be bringing prosperity to your families and to your beautiful country. Honour and long life to His Excellency Mohammed Said Pasha!' The workers, as recorded in his presidential report, received his words with loud cries of acclaim, then 'set to work with ardour, under the direction of the contractor and the foreman of works'. The digging of the Suez Canal had begun.

This historic ceremony was not graced by the presence either of Egyptian or of foreign Consular officials. The news of it first broke in Lesseps's own French-language newspaper, devoted to the affairs of the Canal. It caused a flurry in Consular circles, and a ferment abroad. Here, it was evident, was no mere formal gesture. It was a challenge and a calculated risk by Lesseps, and for many months after it the fate of the Canal, thus inaugurated, was to hang in the balance. Lesseps had thrown down the gauntlet in earnest. For his presidential report, unanimously endorsed by the Company's Commissioners for submission to the Viceroy, embodied a specific programme—quite apart from the plans for a fresh water canal—for the immediate construction of a lighthouse at Port Said, workshops, quarries, huts, and the digging of a *rigole,* or service channel, for the transport of fresh water and materials along the whole line of the projected ship canal, from Lake Menzaleh to Suez.

So the British Consul-General reported to London. He added that an official denial of the Viceroy's consent to these works had not appeared in the Press, as promised, and expressed the opinion that Said, though he might not actively support this programme, was a man of too weak a character to resist Lesseps 'unless he receives from Constantinople very clear and peremptory orders, or from Her Majesty's Government a very distinct declaration of its wishes and intentions'.

Strong instructions were thus sent to Sir Henry Bulwer, who had succeeded Lord Stratford as Ambassador in Constantinople, that 'the Porte should give positive orders to stop a work which was a political and private piece of swindling'. The result was a letter from the Grand Vizier to the Viceroy, pointing out that such works, without the sanction of an imperial firman, were an infringement of the Sultan's rights.

Thus, throughout the hot Egyptian summer, the fortunes of the Canal ebbed this way and that between opposing diplomatic forces—the hostility of Britain, actively fighting to stop it and to stop the Porte from permitting it; the armed neutrality of France, passively disposed not to stop it, liable at a given moment to support it, but at this moment presenting towards it a front which appeared hostile. Looking from one to the other were the two faces of Said, wanting the Canal but wanting other things too, zealous only to displease neither side too much for too long.

Straight between them—and ahead of them—marched Lesseps, now in his early fifties, foursquare, resolute, and alert, sometimes discouraged but always defiant, planning, persuading, publicizing, calculating and manœuvring with the shrewdness of long tactical experience. A man of action above all, he was profoundly practical. Now that the pick-axe had broken the ground, he would not deviate a step from the resolve that the march of his team across the sands of the Isthmus should continue without interruption. The works now begun must not stop for an instant, whatever the obstacles placed in their way. The strength of his *fait accompli* lay in the Canal's concrete progress, and though its momentum might have to slow down it must at no cost be arrested. Were this to happen, all might be lost. Otherwise, his enemies might well wake up to find that the Suez Canal had been dug while they protested and argued against it.

To the Porte's call to order, the Viceroy's reaction was two-fold. The Grand Vizier's letter had been couched in roundabout and even hypothetical terms. In his reply (as he quoted it to the British Consul-General), Said asked for clear instructions as to whether the Canal should or should not be made. If the works were to be stopped, then he requested the Porte to take action itself, notifying the fact to the Ambassadors concerned, that they might instruct their Consuls to withhold support from Lesseps and withdraw their respective nationals now at work on the site. Meanwhile the Viceroy himself, through his Minister of Foreign affairs, Sherif Pasha, issued a circular to the Consular Corps, requiring them to do so.

To avoid confronting Lesseps in person, he delegated to Sherif the task of instructing him to put a stop to the works, which could no longer be defined as preparatory studies. Against this Lesseps protested strongly, circulating his protest to the Consular Corps, but obtaining support only from the Spanish Consul-General. Recapitulating his various communications with the Viceroy, he quoted chapter and verse in an endeavour to prove that the preparatory works, as he still preferred to call them, had in fact been authorized by His Highness in some detail. He made it clear that he was carrying them out, not on his own account, but on that of the Company and its Commissioners. As its President he refused to have further communication with Sherif, and insisted on dealing, as he had done hitherto, with the Viceroy alone—though at present the Viceroy was disregarding his letters and evading his requests for an audience.

In the Isthmus the prohibition rebounded against the Porte, provoking the Canal workers to instant defiance. Typical of this was a station where almost nothing had been done to construct living quarters, usually huts built of wood. Now (as a British observer later recalled), 'it serves to show the degree of French enthusiasm in this undertaking, and the thorough implication of national jealousy, that, on receipt of this message, every member of the encampment, from the first engineer to the lowest dependent, began, by sheer manual labour, to establish themselves more securely by erecting these houses of stone'. In response to a British protest against such disregard of orders, the Viceroy retorted that he had done all he could safely do, in ordering the withdrawal of the Egyptian workmen, and that the rest was the affair of the Consuls. There were Frenchmen at Port Said who had threatened to fire on any of his officers who obstructed them. If he were to send troops a collision would inevitably occur, some Frenchmen might be killed and then the French Consul-General 'would come down on him for something more than an indemnity'.

The Viceroy was contradictory in his actions, changing his orders from day to day, liable to give one order in public and another in private. But despite his caprices—and, as Lesseps suspected, British Consular intrigues—the works proceeded, with a small but for present purposes adequate labour force. Obstruction by local authorities none the less continued, interfering with the supply of labour and provisions. Thus eventually Lesseps, disturbed at the course of events, felt it necessary to depart from his normal paths of discretion, and to send the Viceroy a frank letter of warning.

He confessed to surprise that Said should yield to the pressure of those enemies of the Canal whom he had himself once defined as the enemies of his dynasty and of Egypt, by disowning the mandate granted to Lesseps to form the Company and to ensure its regular operation. 'Your Highness', he wrote bluntly, 'is no longer free thus to default on his sacred engagements, contracted in face of the civilized world; nor am I myself free to delay, by shifts or civilities, their complete fulfilment.' The Company, he continued, had been irrevocably and legally constituted, and must be left free to carry out its engagements.

To seek abruptly to prevent it, after seven months of existence, from exercising its legitimate rights, would be to compromise most gravely your own responsibility. . . . Your Highness is responsible for nothing, if the Company . . . follows, without interference on your part, the prudent and conciliatory course which it has adopted. Your Highness is responsible for everything, if the shareholders . . . claim damages from you for the non-fulfilment of these engagements by virtue of which they formed themselves into a company. My colleagues and I are persuaded of the dangers which this situation presents for the glory and the interests of Your Highness. But we are likewise persuaded of the rigorous duties which this situation imposes upon us.

The Viceroy expressed himself 'cut to the quick' by the letter, and protested strongly against it to the French Consul-General. Its contents were made public, and provided new fuel for the Canal Company's enemies. Lesseps's object in issuing his threat was not in fact to extract an indemnity from the Viceroy, but to awaken in him a salutary sense of the dangers inherent in his present attitude. From his knowledge of Said's character, he judged that this was a moment to play on his fears rather than his vanity.

In British circles there was now open talk of the Company's coming liquidation. 'It is generally felt', read a Consular despatch, 'that the sooner some arrangement can be come to for compensating the "Compagnie Universelle" and quietly winding it up, the better will it be for Egypt and the less danger will there be of grave political complications, which the existence of the Company may at any time induce.' The bankers were discussing ways and means of achieving this, with the least possible loss to the Egyptian Treasury. Securities, it was agreed, had fallen in value, Canal shares were at a discount, shareholders would be anxious to take back their money if no serious loss were involved; and the Viceroy might gladly make a moderate sacrifice 'to be rid for ever of the worst speculation on which he has entered, whether with a view to his own interests or those of the country which he governs'. The British Consul-General strongly urged him to take this course, suggesting that the French shareholders might be easier to manage than Lesseps, and that 'the Governments in Europe would more readily come to an understanding than their representatives in Egypt'.

More disquieting to Lesseps was a loss of confidence not merely among shareholders but among his own partners in Paris. 'What is Lesseps up to in Egypt?', enquired the Duc d'Albuféra, the Vice-President of his Company. To the Council he complained: 'Always very great efforts . . . always very small results . . . We pretend that we are the masters in Egypt. . . . We know all too well we are not. . . . Only one force can overcome our obstacles, the force of the French Government acting with all its power. . . . Lesseps is moving but he isn't making the Canal. . . .

Until we get decisive action from the French Government all that we are doing means nothing.'

In his capacity as Vice-President, he wrote to Walewski, requesting such action through the Emperor, but received only a negative reply on the familiar grounds that the Porte had not ratified the Concession. Even Lesseps's most devoted friends were beginning to lose heart. 'Ought we not to accept our situation,' enquired one, 'and give up fighting openly against the Porte?'[1] 'It was all very well to say', wrote another, 'we shall maintain our guarantee of our shareholders' rights, but with what? If in a few days' time our Emperor will not support us energetically, our future course will be difficult.'[2]

Lesseps was well aware of the need for more positive official backing from France, and thus for his own presence in Paris as soon as the situation allowed him to leave the works in the Isthmus to the men on the spot. He had written to his brother in Paris, for the ear of Walewski, hoping that he would not be left to hold the fort on his own. He mistrusted the French Consul-General, Sabatier—as it turned out with reason.

Since the start of the works in the Isthmus, Sabatier had been expressing to Paris misgivings about the actions of Lesseps, and their repercussions in Egypt. In the course of a long despatch he assessed the situation in terms of his own opinions. Firstly, he had asked himself, had the Suez Canal Company been regularly and legally organized? To this his answer was No, it had not complied with the formalities required by the law with regard to the full subscription of capital; it was at fault in raising this capital in the knowledge that the works on the Canal could not legally start without an authorization from the Porte, and that this had not yet been obtained. Hence the Viceroy, though he had authorized Lesseps to organize the Company, was within his rights in disclaiming responsibility for its actions. He was justified in opposing its operations in the Isthmus, having authorized preparatory studies, but specifically not preparatory works, on which the Company was now openly engaged. Against these the Viceroy was entitled to intervene, and for the Company to

[1] Amedée de Chancel.   [2] Victor Delamalle.

threaten, as it had done, to oppose his intervention by force was nothing more nor less than an act of madness.

Sabatier however had taken no action to withdraw the French employees at work in the Isthmus, and he could only hope that, when tempers had cooled, Lesseps and his associates would see the danger of their position and no longer persist 'in the detestable course on which they seem to be embarking'. Meanwhile he would carry on with his ungrateful task of attempting to reconcile the two parties.

Walewski, in an equivocal reply, approved this course, provided that it did not compromise the individual interests of the Company's French employees. He reminded Sabatier, however, that a large part of its capital belonged to French shareholders and that it was necessary, without interference on the political issue, to safeguard their interests.

His words, though ambiguously phrased, were significant. These interests were now such that France could hardly maintain for an indefinite period a front of official indifference towards the Canal scheme. Now that, in spite of her attitude, it had progressed so far as the formation of a Company involving 21,000 French investors—worthy middle-class citizens of a middle-class Empire—she could not afford to see it fail as a result of hostility from Britain. National prestige must commit her to a more positive support for the Canal. The time was ripe for a reorientation of policy at the Quai d'Orsay.

CHAPTER IX

An Appeal to the Emperor

By the end of July 1859, Lesseps judged that the situation in the Isthmus was, for the moment, stable enough to permit of his departure for Paris. Before leaving he had a reconciliation with the Viceroy, who received him with some of his former familiarity and, as an apparent proof of friendship, entrusted him with the care of his young son, Prince Toussoun, who was to stay for a while in France. Lesseps found Said far removed from any intention to abandon the Canal or the Company's interests, but now assured him that, in the event of its abandonment he would claim no indemnity. He would simply restore to the Viceroy the rights of the Concession and those of the Company, and hand over to him its lands, installations and materials in the Isthmus, against his reimbursement, for the shareholders, of the Company's expenses. To this Said verbally agreed. Thus Lesseps left Egypt, as he put it, 'with the honours of war'.

Arriving in France he sought to achieve the honours of peace. First he wrote an astute letter to the Viceroy, absolving him from any responsibility for the present dispute, and placing the fate of the Canal squarely on the shoulders of Britain. To her opposition there was at present no counterweight. If Britain were to triumph, he undertook to relieve the Viceroy from all political embarrassment, on the terms they had discussed before his departure. But if, as he hoped and believed, British opposition was defeated, he reminded the Viceroy that he had reserved for him the foreign shares of the Company, adding them to his own subscription and so giving him a valuable investment.

After talking in a similar sense to Walewski, Lesseps turned his pressure on the Emperor himself. In a note to him he explained that the operations of the Canal Company were being hampered, without counterweight, by the political intervention of British agents in Constantinople and Alexandria. The Company, despite British-inspired demonstrations against it, was maintaining its rights and continuing its works in the Isthmus. It was to call a General Meeting of its shareholders in November next. If by then British political threats still prevailed, Lesseps, as President of the Company, would feel obliged, in the political interests of the Viceroy, to propose at this meeting the liquidation of the Company, the reimbursement of the shareholders, and the cession to the Viceroy of the Company's property in the Isthmus. Though such a solution would damage French prestige and influence, not only in the East but in the world in general, it seemed inevitable unless measures were taken very soon to counteract British pressure.

Lesseps thus felt obliged to solicit the protection of the French Government against the British. This could be afforded in two alternative ways. Either let the Emperor, through his Ambassador in Constantinople, request from the Porte, unofficially if need be, the Sultan's firman—a request which, coming from this high source, would surely be granted. Or let the Company petition the Emperor to request from the British Government an explanation of its motives in opposing the Canal. In any ensuing negotiations the Emperor should be able to count on strong international support.

Lesseps then retired for a while to La Chénaie, where he was able for some weeks to give the young Egyptian Prince a taste of French country life. But early in the autumn his rest was disturbed, and a crisis precipitated, by a storm which broke over Egypt from Constantinople. The Porte, since its previous injunction had failed to stop the works in the Isthmus, had sent as an official emissary to Egypt its Minister of Justice, Mukhtar Bey, with a stronger letter from the Grand Vizier. This ordered the Viceroy to suspend them without further delay. It was pointed out to him that the piercing of the Isthmus might make Egypt the theatre of conflict in any future European War, leading not only to its separation from Turkey but to the removal of the Mohammed Ali dynasty.

Sherif Pasha immediately summoned the Consular Corps and, after reading aloud the Vizierial Letter, instructed them for the last time to remove their nationals from the Isthmus. This time the instruction was based on legal grounds and accompanied by a threat of coercion if necessary by armed force, in which the Consuls were expected to collaborate. There was a brief pause. Then a voice spoke up, acquiescing in the demand. It was the voice of Sabatier, the French Consul-General. His agreement, as doyen of the Consular Corps, was echoed by his colleagues without a word of dissent or protest. Sabatier, who over the past months had come to see the Canal as a lost cause, at once suited his actions to his words by issuing an order, through his Vice-Consul in Damietta, that all French employees should leave the Canal zone.

This was a moment of life or death for the Suez Canal. No one, it seemed, could now save it but the Emperor Napoleon III in person. Lesseps, who had meanwhile been paving the way with his 'guardian angel', the Empress, secured an audience with the Emperor at St. Cloud on October 23, 1859, for himself and a delegation from the Company's Council. The Emperor at once took the initiative.

'How is it, M. de Lesseps,' he asked, 'that so many people are against your enterprise?'

'Sire,' he replied, 'it is because everyone believes that Your Majesty does not wish to support us.'

After reflecting for a moment, slowly twirling his moustaches, the Emperor replied: 'Well, don't worry. You can count on my support and protection.' Referring to the attitude of Britain and to a recent exchange of notes on the subject with London, he added: 'It's a squall, we must trim our sails to it.'

Lesseps then explained to him that, if the Company were to be saved, he must be able to tell his shareholders that diplomatic negotiations were starting. For this reason it would be necessary to postpone the General Meeting, fixed for the middle of November. Only thus could liquidation be avoided. The Emperor agreed. He also implied that negotiations would be initiated by France.

After the other delegates had taken their leave, the Emperor detained Lesseps for a moment, with his Vice-President, and asked amiably: 'What do you think should be done at this moment?'

To this Lesseps replied: 'Sire, a change of residence for the

French Consul-General, a diplomat of considerable capacity, who could well occupy another post.'

'If that is all,' said the Emperor, 'it is easy enough. Tell Walewski.'

Thus Sabatier's career was to be sacrificed to the equivocal policy of the Quai d'Orsay, itself a mirror of the Emperor's own indecisions. In a long, stern despatch from Walewski, he was reminded that, whereas he had been warned not to commit his Government to support of the Company, it had been made equally clear to him that he must not oppose it, especially in view of the fact, stressed in his most recent instructions, that the greater part of its capital was French. He was therefore at fault in failing to uphold the Company's rights at the Consular meeting. Instead of assenting to the Egyptian Government's demands, he should have refused to accept them, requested their modification, or delayed a reply, pending instructions from Paris. He was instructed to explain his actions, and meanwhile to request the Viceroy, in the name of the Emperor's Government, to order that the Company's installations in the Isthmus be left undisturbed, pending a pronouncement by the Porte in the light of its legitimate rights under the Viceroy's Concession. He was also to ensure the protection, through the Consulate-General, of the Company's employees.

To this Sabatier replied at length that he had remained from the start strictly faithful to his instructions, and that these had at no time changed. The Foreign Minister's most recent despatch, calling his attention to the French interests involved in the enterprise, had approved his previous conduct and contained no new instructions, beyond commending these interests to his support, 'without interference in the political issue'. But where did politics begin and end? There was only one issue—whether the Isthmus should be pierced or not. And on this the Foreign Minister had given him no word of enlightenment. If, however, the Emperor's Government persisted in its censure, he could only ask to be relieved of his post. Sabatier's resignation was accepted and he retired from the service with Ministerial rank. His recall greatly enhanced the prestige of Lesseps and his Company in Egypt.

The immediate crisis was eased, and a new landmark was reached in the progress of the Suez Canal. The Emperor's intervention, nebulous as it was in its terms, achieved no striking transformation. The way ahead was still littered with obstructions and beset with hazards. The British Government was to remain as implacable, the Porte as intractable as ever. But at least the French Government had at last committed itself to a degree of overt support for the Company and its project of piercing the Isthmus. Lesseps, in his continuing battles with Britain, could at last claim an ally on his own home ground. For his cause now admittedly involved the honour of France and the national interests of Frenchmen.

*   *   *

Until the attitude of Paris thus hardened, London was congratulating itself that the battle against the Suez Canal had been won. This unfortunate but ephemeral project seemed dead, and in the process of burial; the works had been stopped and the Company would soon be wound up; the file was closed. Lord Cowley confidently hoped that 'no further correspondence would occur respecting this vexed question'. Hence he had been disconcerted by a new approach from Walewski, sounding out the British Government's attitude afresh. For there had been a change of Government, with Lord Palmerston back in power as Prime Minister and, as his Foreign Secretary, Lord John Russell, who in opposition had supported the Canal, on general Liberal principles, against Disraeli in Parliament.

Walewski informed Cowley that the Company had addressed a memorial to the Emperor stating that the only obstacle to its success was the hostility of Britain, and requesting his intervention in its favour. He explained to the Ambassador that his Government, though favouring the scheme, had discouraged it in deference to the wishes of Britain, and that Lesseps had been asked several times to abandon it. If, now that Her Majesty's Ministers had changed, the British Government had modified its opinion, the Emperor would hear of it with 'unfeigned satisfaction'. If so Walewski proposed that the two Governments should either make a joint approach to the Porte, to obtain the firman, or should abstain from further opposition. Otherwise Walewski himself considered that the Company must resign itself to a liquidation of its affairs.

Inevitably London returned the familiar negative answer. Opposition would not be relaxed and the British Government saw no advantage in further discussion. At this rebuff Walewski changed his ground, declaring that if Britain would not agree to joint non-intervention at Constantinople, a clash between British and French influence on the Porte must follow, since the French Government could not possibly abandon the interests of the Company, which had claimed its protection.

Thus, following Lesseps's audience with the Emperor, new instructions were sent to Thouvenel, the French Ambassador to the Porte, as to the policy he must now pursue. In one sense, these were as moderate as before: all friction with the British Ambassador must be avoided. On the other hand, they showed a significant change of emphasis. The French Government must protect French interests, and a continued postponement of the Porte's sanction would condemn the work of Lesseps, in which the French nation was so largely interested. Thouvenel was instructed to seek an agreement on the subject between France, Britain and Turkey, and meanwhile an assurance that the preparatory works in the Isthmus would be allowed to continue.

This new French policy, involving as it did for the first time official support of Lesseps and the Canal scheme, created a commotion among the Sultan's Council of Ministers. It was accompanied by the arrival in Constantinople of Lesseps, this time describing himself not merely as President of the Canal Company but as a representative of the Emperor. The Porte—a Government, as Thouvenel harshly described it, 'completely inert and devoid of initiative in its foreign policy as of capacity in its internal policy'—was now in a quandary. It was a Government virtually without a will of its own. From the start it had been concerned relatively little with the merits or otherwise of the Canal scheme as such—with arguments, for example, as to whether it would benefit the communications of the Ottoman Empire. Its concern was rather with the danger implicit in the choice of a policy of any kind towards it—that of antagonizing one or other or both of the two powers, Britain and France, on whom Turkey relied for her security.

As long as France remained neutral on the Canal issue, it had been easy enough to please Britain by the negative policy of taking no action. But now each had become the open antagonist of the other. France was demanding positive action—the ratification of the Concession—which Britain opposed. In vain would the British and French Ambassadors urge the Porte, in their respective interests, to display its independence and assert its own wishes. It had no independence and barely desired it, no wishes but to find a formula which would absolve it from doing anything, one way or the other, and would displease neither side.

The shift in French policy made a direct choice by the Porte between sanction and prohibition more remote than ever before. In the ensuing diplomatic exchanges it became the tactic of the Ambassadors, in their struggle for influence, to suggest to it such formulae as might, if they led to no immediate conclusion, suit their ultimate purposes best. And this process, in the long run, was to work against Britain and in favour of the Canal and of France.

The French formula was to remove the question of the Canal into a wider sphere, to raise it from the national on to the international plane, where it properly belonged. Thouvenel advised Fuad Pasha, the Turkish Foreign Minister, that only thus could the Porte retain some dignity in the eyes of Europe. The Austrian Internuncio, friendly once more now that the war between Austria and France was over, gave similar advice; the Spanish Minister was co-operative; the French Government sounded out the Governments of Russia and Prussia on this policy with some success; and it was suggested to the Porte that it should address a note to the powers, asking them to determine guarantees considered necessary for a Canal which would link the two seas.

The Emperor Napoleon, who saw in congresses a sovereign remedy for all international ills, was now planning a European Congress to follow his Peace Treaty and settle, with other matters, the reorganization and reconstruction of Italy; and it was thought that the question of the Suez Canal might be placed on its agenda. Just so had Lesseps, on Metternich's advice, sought to place it on the agenda of the Congress of Paris, four years earlier, after the end of the Crimean War.

The Turks were at first inclined to agree to such an international appeal. But it could only provoke strong disapproval from Britain. Palmerston, in his old age, abominated all congresses in any case. In this particular case of an international conference on the Suez Canal, it was obvious that Britain would find herself in a minority of one, with public opinion in Europe against her. She thus refused to participate in any such joint diplomatic approach by the Porte. The London Press (on December 16, 1859) took a strong but conflicting line on the subject:

If [declared *The Times*] the French or any other people are bent upon sinking their money in the sand. . . . If France, Austria, Prussia, Russia and Sardinia are so decidedly committed to this unpromising enterprise as to concert measures for investing it with the patronage of the Sultan . . . we need go no farther . . . in our opposition to the scheme. In fact, as we are firmly convinced that the works can scarcely be executed, and can certainly never be maintained, we cannot profess to conceive any apprehension of an impossible result. While as to the money, not much of that will come out of British pockets. . . . If, however, contrary to all probabilities, the project should be actually realised, we can only say that the Canal will be so far a British Canal that it will be traversed by British ships, devoted to British traffic, and maintained by British tolls.

The *Morning Herald* called the Canal 'one of those magnificent "ideas" of which France claims the monopoly', and compared it to the idea behind her recent war in Italy. This international approach to the Porte, and the refusal of the British Ambassador to participate, would encourage 'the very general belief that the opening of the Suez Canal will be a great blow to English interests in the East, and consequently an enormous benefit to the Continent'.

The *Daily News* likewise saw the approach as a move by

France to represent Britain as 'under the ban of Europe, totally isolated upon a great question, and opposed to the progress of civilization and intercommunication, from superannuated fear and jealousy about the integrity of Turkey, and the danger of the Mediterranean ceasing to be a *mare clausum*. The object, no doubt, of bringing forward the Suez scheme at the present moment is to show how England is capable of taking a mean and selfish ground, unsupported by any other power.' In fact, 'we aim at the great objects of keeping Turkey from Russian annexation, and Italy from either French or Austrian. Let us keep our efforts and our will fixed upon great purposes, and not be carried away by the petty fancy of securing the Egyptian desert from French speculators. Deserts will take care of themselves. Why make enemies for purposes which soil and climate can always of themselves accomplish?' Lord Palmerston's *Morning Post* bluntly declared that England had no interest 'in creating for the especial benefit of France an Egyptian Dardanelles or an Egyptian Gibraltar'.

The British Ambassador to the Porte was now Sir Henry Lytton Bulwer, Lord Stratford de Redcliffe's successor, who had served as a young Secretary in Constantinople some twenty years earlier.[1] Bulwer and Lesseps had been diplomatic colleagues in Madrid and had shared those liberal views which had led to Bulwer's dismissal as Minister by the Spanish military dictator, Narvaez, in 1848. Now, in Constantinople, they were diplomatic opponents. Lesseps still looked upon him as a personal friend. But Lesseps's arrival, while the Ambassador lay in bed 'with a fever', seemed to have acted upon him like 'a good dose of quinine'. For he rose and left for the country the very next morning.

'All the better!' exclaimed Lesseps. 'Shock can give birth to light.' On the issue of the Canal Bulwer proved an adversary different from Stratford but perhaps more insidious, whose technique was to disarm rather than to intimidate, and who concealed, beneath an easy and even languid conversational manner, determined views and shrewd perceptions. A French colleague—

---

[1] Brother of the first Lord Lytton, the novelist. Ambassador in Constantinople 1858–1865. Afterwards created Lord Dalling and Bulwer.

Beauval, the Consul-General in Egypt—wrote of him: 'He imposes himself by his grand manner, which none the less does not exclude either benevolence or charm. His mind is brilliant and backed by education, by the force of experience in public affairs, and by a singular skill in defending dubious theses with an appearance of honesty and directness of approach.' These talents were now applied to negotiation with the Ministers of the Sultan's Divan.

The Divan's present discussions continued interminably. Meeting day after day, the Ministers smoked endless pipes, reached no decisions, and provoked Lesseps to the Latin quotation, *'Consternebantur Constantinopolitani'*. Their 'anguish' was such that, as Thouvenel replied, they would soon be requiring a *'rigole de service'*—a service canal—to drain away their tears. Bulwer advertised the fact that in his Ministerial interviews his aim was to demolish each evening what Thouvenel had built up in the morning. He soon established an influence, shrewdly discerning and exploiting a rift of opinion between two groups of Ministers.

On the one hand was Fuad Pasha, the Foreign Minister, taking the view that if the matter of the Canal were officially brought forward, with the French Embassy's backing, the Porte, as he put it, 'will either have to accept M. Lesseps's conditions, which would be hostile to England, or to reject them, which would be hostile to France. By leaving Europe to decide the matter we would escape from this which is our first difficulty at once, and if Europe does not agree, we may yet escape from our second.' Bulwer opposed this view, on the grounds that it would make France 'the arbiter, in a European Congress, of a great Turkish question in which the views of France and England differ'. It would take the matter out of the hands of the Turks, and needless to say, of our own.

He preferred the attitude of Kiprisli Pasha, the Grand Vizier, who was inclined to argue: 'If we invite Europe to treat this question as rather of European than Turkish importance, we pass it out of our own hands, and must eventually receive the law from Europe as a matter of course. We also deal with disadvantage with Europe when we deal with it collectively; and we shall, in the present instance, bring England into conflict, less friendly to us, but who will have a great majority against her opinion.' He would sooner treat separately with each of the interested powers: 'The question is still in that manner within our own grasp.' Bulwer upheld this course. He did not see how it could 'lead fairly to any complaint on the part of France'; and it was in fact to serve as a basis for the policy which he was finally to persuade the Porte to adopt.

Meanwhile, official instructions were despatched to Bulwer by Lord John Russell as to the line he should take. In rejecting the international approach and approving the views of the Grand Vizier, Lord John introduced a phrase which implied some slight modification of policy—or at least of tactics. He referred to the Canal scheme as a 'secondary interest' for Britain, declaring that she could not undertake to resist it if Turkey, for whom it was a 'primary interest' refused to do so.

Here was a different tone of voice from the earlier and more belligerent despatches of Palmerston. Recent events seemed to have shown the British Government that it could no longer easily oppose a start on the Canal's construction. But it could still effectively delay and impede its progress. This, rather than out-right prevention, was from now onwards its guiding principle. For these purposes Bulwer furnished it with several obstructive weapons, of which two at least—the curtailment of the Company's labour and of its grants of land—were to prove in part effective. Obtaining the Porte's verbal agreement, he wrote to London:

I got accepted, as the basis of all future proceedings, first that the Turks should not give their consent to the Canal without the project being first stripped of its features of colonization and forced labour on which M. de Lesseps had made most of his profits depend.

Secondly, that such consent should be withheld until the advantages and expenses of establishing the said Canal had been, as a preliminary measure, fairly and fully gone into.

Thirdly, that whatever fortunes should be made upon or with respect to the work in question should be kept in the hands of the Sultan, and fourthly that the guarantee of England which implies her assent should be held as a *sine qua non* of the whole undertaking.

The Turks in their turn, seeking to shift from their own shoulders the full responsibility of a final decision, abandoned the project of an appeal to the powers in general, but adhered to that of obtaining prior agreement from Britain and France, the two maritime powers directly concerned. The Porte sent a note to the two Governments, through its Ambassadors in London and Paris, requesting consideration of three basic questions concerning the Canal. It sought assurances that the Canal would in no way endanger or prejudice the Ottoman Empire; safeguards for the Empire's right of sovereignty in Egypt, and arrangements for the Canal's defence; guarantees against the effects on the Empire of any conflict between the powers which might, in the future, arise out of the Canal. On these matters it sought joint agreement between the two powers. Until it could be assured of this, the Porte would not decide whether to give or withhold its consent to the project.

No such agreement was reached; nor was it seriously contemplated. No assurances were given by Britain in relation to the three points put forward; only long arguments, none of them new, against giving them, and a categorical refusal to give a British guarantee. The Porte did nothing either to grant the firman or to refuse it. The time-honoured principle of delay, the negative approach to the problem, was accepted once more as an instrument of Ottoman policy.

Only the makers of the Canal themselves did not delay. The Viceroy had been instructed to stop further preliminary works until a decision was reached by the Porte. But he turned a blind eye to this instruction and gave tacit approval for the works to continue. The *fait accompli* thus proceeded without serious loss of momentum. Lesseps, by his resolution and resource, had weathered the crucial year 1859. Now, from 1860 onwards, preparatory traces of a Canal began, very slowly but reasonably surely, to take shape across the sands of the Isthmus of Suez.

CHAPTER X

A Canal takes shape

The first Annual General Meeting of the Suez Canal Company, postponed from November, was held in Paris in May 1860. The report of it, as published in Alexandria, was described by Bulwer to Lord John Russell as 'a plausible perversion of fact'. But the shareholders seemed satisfied. No awkward questions were raised, and 'full of hope and enthusiasm', they voted unanimous acceptance of their President's proposals.

One of the main problems now confronting Lesseps was that of the unsubscribed shares, allotted to the Viceroy but not yet taken up. This had to be tackled before the next General Meeting. Mohammed Said had little sense of the meaning or value of money. His personal extravagance was proverbial. His generous habits, coupled with his desire to lead an easy life and to save himself the trouble of irksome decisions, put him at the mercy of all who chose to make demands upon him. Exploiters would raid the Viceregal Treasury without scruple and with habitual success, seeking concessions for dubious projects, making legal claims against him for loss or indemnity, charging exorbitant interest rates for loans, lining their pockets in a hundred artful ways. Attributed to him is a dry (and perhaps apocryphal) remark, when in conference with a European business man. Interrupting their talk for a moment, he ordered his servant to close the window, explaining: 'If this gentleman catches cold it will cost me £10,000.' Friends could always be sure of Said's generosity, in terms of lavish gifts or permanent loans without interest; of expansive promises of support made without heed for the consequences.

But Said, with his volatile temperament, would fly suddenly from an excess of extravagance to an excess of caution. Growing alarmed at the burden of his expenditure, he would seek means of evading his more ruinous commitments—and at the same time, in pursuit of the line of least resistance, would contract others as ruinous.

In backing the Suez Canal, he had been animated by a combination of political ambition and personal friendship. Lesseps well knew how to play up to both impulses when it came to finance. In the atmosphere which he created between them, the Viceroy had given him airy and spontaneous undertakings without counting the possible ultimate cost. And now, after a respite of eighteen months, the day of reckoning was upon him, in terms of the Company's need for hard cash and for the regularization of its status.

His original subscription was for 64,000 shares at a hundred francs each on the first call. This obligation he had accepted, and towards its discharge he had already paid out some 2,500,000 francs. But on his subsequent obligation to take up and pay for the remaining 113,642 shares, he had always been evasive and restive. Lesseps, in an account presented on behalf of the Company, sought to hold him to this and thus to a total outstanding liability of some 15 million francs, on the grounds of the unconfirmed verbal agreement between them. Said had received independent legal opinion from Paris that the Company had been irregularly constituted, but that his own liability was no different from that of an ordinary shareholder. Thus he was advised to avoid giving the Company's agents a chance to come down on him personally for further liabilities.

Said was still reluctant to take up the remaining shares; and Robert Colquhoun, the new British Consul-General, was zealous in his efforts to play on his fears and dissuade him from doing so. At first Colquhoun found him responsive. In an interview (as reported to London) the Viceroy 'said in his usual gay manner that the Canal Company had endeavoured to *fasten on his shoulders* the whole of the shares remaining unclaimed and which had been assigned to various nations'. The result, in the Viceroy's words, was that 'I find myself, by I don't know what conjuring trick, down for nearly a hundred million francs'. (The sum which he chose to quote was that of the total value of the shares, at the final figure of five hundred francs each, not on the present first call of a hundred francs each.) When Colquhoun asked whether he had accepted this charge, Said laughed and replied: 'I am not such a fool.'

But Lesseps was now back in Egypt, and quick to relieve his master's fears at the expense of their common enemy's pressure. He proposed to him, and confirmed with his Finance Minister, an arrangement by which the Viceroy should make no initial payment for two years. Thereafter, the 15 millions due would be paid off by instalments over a period of three years. At the welcome prospect of this period of grace, he exclaimed: 'Who can tell what we may all be in two years?' When Colquhoun, spending an evening with him at his seaside palace near Alexandria, continued to lecture him on the error of his ways—his subservience to the Canal Company, his improvidence and the state of his finances in general—the Viceroy, 'after a pause in which he was evidently labouring to contain his feelings and to repress anything like anger', said:

'Mr. Colquhoun, I think I have before told you that this scheme is a child of my own begetting—a scheme I have looked upon since I became Viceroy as the one great act of my life, which is to hand me down to posterity as a worthy son of my father. I cannot but consider the junction of the two seas as an undertaking which will immortalize him under whose auspices it is carried out. Had I refused to take this large part of the shares in the Company, I should have seen them thrown into the market; the natural consequence would have been a ruinous depreciation, perhaps a complete break-up of the scheme.'

When the Consul-General persisted in his admonitions, pointing out that if there were any glory to be achieved from this unremunerative scheme, it would fall to Lesseps at the Viceroy's expense, Said 'abruptly broke off the conversation and during the rest of the evening was frequently absorbed in anxious thought'. Colquhoun's verdict on the situation, as reported to London, was that the Viceroy was 'entering upon engagements which must inevitably hurry him on to his complete ruin'.

An agreement was reached between the Company and the Viceroy's Finance Minister on August 6, 1860, that payment on the first call for the Viceroy's shares, amounting to 15,248,042 francs, should be made, by monthly instalments, starting in January 1863 and ending in December 1866. By this agreement the uncalled balance of 400 francs per share would be paid by similar instalments between January 1867 and January 1875.

This agreement, however, was never signed by the Viceroy, nor by any of his Ministers. Thus the financial situation remained much as before, apart from the fact that the Company could claim that it was regularly constituted according to law. Nor did Lesseps succeed in subsequent attempts, under pressure from the Company, to induce the Viceroy to pay cash for the total amount of his shares by an issue of thirty-year Government bonds.

*   *   *

In April 1860, a year after the start of the preparatory works of construction and the stroke of the first pick-axe, Colquhoun had sent a qualified observer to the Canal site to report on its progress. At present, apart from the Sweet Water Canal, the work was confined to the northern end of the Isthmus—to the building of the harbour of Port Said and the digging of the first reach of the service canal—the *rigole*—across Lake Menzaleh and so to Lake Timsah in the centre. But a string of eleven stations had been provisionally established along the whole line of the projected Canal, whose course was marked out at intervals by stakes.

Each station was manned by a group of the Company's European employees, living and working in houses built sometimes of stone but more generally of imported timber and local mud-brick or rubble, with roofs made of reeds from Lake Timsah 'interlaced very ingeniously'. Workshops and stores were at hand, together with lime-kilns and brick ovens for the use of the builders. The Arab labourers lived near by in tents or mud-huts. In the larger stations the blocks of buildings, which included a hospital, a canteen and a general store, were already laid out, with boulevards and streets-to-be in the French manner, each named after the

Company's more prominent engineers and shareholders—Rue de Lesseps, Rue Ruyssenaers, Rue Mougel.

One of the stations around Timsah—never in fact to be built—was named Lessepsville; another, soon to be abolished, was named Toussoum, after the Viceroy's young son. Situated to the south of Lake Timsah, it served as a general supply centre for the works of this sector, and the Consular observer found it reasonably well organized. Throughout the night coloured lights were hoisted to guide Arabs bringing in fuel from the desert; throughout the day bells were rung at regular times to mark the shifts of the native workmen. Pains had been taken to lay out some kind of a garden, the soil being frequently washed to remove saline deposits and irrigated by a windmill with water which was drinkable only by camels. Here too was a forge for the manufacture of minor utensils, and a pound containing livestock—sheep, bullocks and innumerable fowls. The whole station was 'subject to diseases of considerable virulence, the most formidable being dysentery'.

Some two hundred Europeans in all were at work in the Isthmus, men from all quarters of Europe 'with difficulty held together by the new hope of gain'. For the present their salaries were 'so low as to bar any hopes of profit sufficient to counterbalance the daily sacrifice of comfort and society', and many would leave, disillusioned, when the first flush of enthusiasm wore off. They resented especially the fact that they were not allowed to beat the Arab labourers, 'the refuse of the native population', whose 'indolence and insolence' provoked them.

On the site of Port Said the morale of the European community was 'at a low ebb in all respects'. 'We live like dogs,' they declared. Supplies were still short, and an engineer complained of his employers: 'They give me men, and no pick-axes; another time they give me pick-axes without men; and, once or twice, when they have succeeded in giving me both at the same time, they have found it impossible to send provisions.' For evening entertainment their only resort was a kind of 'cabaret', where 'jealous international quarrels' would often break out, mostly between the French and the other main European contingent, the Austrians.

A stretch of the Canal is hollowed out

The men who have hollowed it

The cutting of el Guisr

The cutting of Chalouf

'These drunken discords', the observer reported, 'sometimes proceed very far; and knives are drawn freely in a place where no police or other controlling agency is known.' Altogether there was 'nothing in the natural features of the place which could suggest such a destiny as that intended for it by the Canalists'. The jetty 'shook and trembled perceptibly under slight pressure of wind and wave', and a shallow, sand-choked ditch from the jetty to the lake, 'seems designed only to suggest that a Canal will ultimately exist here'.

Lesseps, with the eye of faith and the imagination of a Frenchman, saw these beginnings through rosier spectacles. To him the 'ditch' was already his own infant, the Suez Canal, which would grow up to unite the two seas. On the Isthmus that summer he enthused in his journal over the dawning Port Said: 'The new town presents a most striking picture, both from the lake and from the sea. Built along the line of the long narrow beach, among sand-dunes which raised it a little above the level of marsh and sea, it delighted the eye with the panorama of its various installations', the pride of which was the dredgers' workshops. The yards were stocked with apparatus and stores unloaded from fifty ships, criss-crossed with rail tracks, and thronged by 'numerous European and native workers' who saluted Lesseps with salvoes of rifle fire.

On Sunday, at an altar erected on the porch of the engineers' chalet, Mass was celebrated before a crowd of employees and their families congregated on the beach. A Lazarist father preached a moving sermon. Inspired by 'the grandiose spectacle before him, he found words to touch the hearts and to encourage the labour of those dedicated to the universal mission of uniting the one sea with the other'. Afterwards, the first child to be born in Port Said was baptized with the names Ferdinand and Said (translated as Felix). At a banquet in the evening, Ferdinand de Lesseps pleaded eloquently for unity among the Company's workers themselves, without distinction of race or creed.

Such social functions were to become a feature of life in the Isthmus, when Lesseps sought to impress influential visitors and to uphold the morale of his men on the spot. One such visitor, that autumn, was a French Consul, Count Colonna-Ceccaldi, who after a night in a barge on Lake Menzaleh, troubled by sea-sickness and the bites of mosquitoes, enjoyed a bathe on the beach, and in the bustle around him found an atmosphere which recalled 'the beginning of one of those new towns in the United States, with that extra touch of sprightliness and grace characteristic of our national spirit'. Inviting the European personnel to a luncheon in the workers' canteen, Lesseps exhorted them: 'Treat the natives well; they are men.' The Count was impressed by this 'little army which follows its chief with absolute confidence, an utter devotion to the idea and to the work, and a conviction of the certainty of future success'.

Now Lesseps, starting on a tour of the northern half of the Isthmus, proceeded across the shallows of Lake Menzaleh into a landscape once watered by a branch of the Nile where 'each village, each ruin is a page of the Holy Bible'. Thus he reached Kantara, the start of the caravan road into Syria. This station, designed to become a large depot of machinery, would be directly linked with Port Said as soon as the dredgers had cut a twenty-five mile service channel through the mud of the lake, and another dredger had been brought overland in sections to be reassembled and start the work from this end of it. Meanwhile a preparatory channel had been excavated by hand. The local labourers, familiar with the mud of their lake and its habits, scooped it up in large handfuls, pressed it against their chests to squeeze out the water, and when dry piled it in lumps, one on top of the other.

From Kantara Lesseps proceeded southwards through a dry swamp which was to become Lake Ballah—an offshoot of Menzaleh—to the sand-dunes of Ferdane, whose excavation was soon to start, and where now, as at Kantara, 'complete well-being, gaiety and good health reigned'. Continuing his journey to the south, he explored the basin of Lake Timsah, twenty feet below sea-level and dried up for centuries past, but now destined within measurable time to be filled by the waters of the Mediterranean. Here, nearly half-way between the two seas, where the sea water canal from the north would meet the Sweet Water Canal from the west, lay the site designed to be the principal inland port of the Isthmus.

Looking across this gaping void of sand and gravel and blackened sun-cracked earth, Lesseps saw as in a mirage the day when it would 'shelter the fleets of all the world'. On its western rim he marked out, with his party, the lines of a town which would be known as Ismailia, after Mohammed Said's successor, Ismail. Southwards they looked down towards the bluish silhouette of the Attaka mountains, which rose above the Gulf of Suez. Between lay the Bitter Lakes, to be linked with Timsah by a stretch of the Maritime Canal and with Suez by another, while the line of the Sweet Water Canal would turn south, to run parallel with both. Such was the plan for the second phase of the works, planned to begin when the northern half had reached a certain stage of completion.

Few technical difficulties stood in the way of the construction of the Suez Canal. It was to be cut in a relatively straight line, through some sixty miles of water and forty miles of land—three large lakes and three stretches of desert between them. The land was in fact a natural depression, parts of which must in the past have held water. Everywhere the soil was easy enough to work, being composed of sea-sand in the north, and gravel and clay in the south, with here and there layers of gypsum (useful for the making of lime), but nowhere, it seemed, more than an insignificant trace of rock.

In the bed of the Bitter Lakes lay a large deposit of crystallized salts, from the days when the Red Sea flowed into them. The land was level enough to require no cuttings except through three ridges or plateaux across the path of the Canal, none higher than forty feet above sea level. These were at el Guisr, between Kantara and Timsah; at the Serapeum, between Timsah and the Bitter Lakes; and at Chalouf, between the Bitter Lakes and Suez. Only one of these cuttings was seriously to tax the skill of the engineers —that of el Guisr where the work was to take longer on account of the problems of excavating soft sand.

Otherwise the technical challenge of the Suez Canal lay less in the piercing of the Isthmus itself than in the construction of its artificial port, on the Mediterranean. Here nature had provided no inlet from the sea to Lake Menzaleh. Nor was there any land but a thin strip of beach, sometimes no more than a hundred yards wide, with the waves washing over it. Moreover the bay, like the lagoon, was exceptionally shallow; ships could only lie three miles out; and any attempt to dredge a navigable channel would be threatened by the prevailing current from the north-west with deposits of silt from the Damietta branch of the Nile.

Port Said thus had to be built up from nothing by man, in defiance of the forces of nature, and with the assistance of no local building materials. It was to be contained within a western breakwater, built to enclose 450 acres of water. First, the pioneers erected a lighthouse some eighty feet high, made of timber and designed for a revolving light, which would be visible fifteen miles offshore. Then they threw out a jetty, built on piles, of which the larger were not merely driven but screwed into the mud and the sand of the sea-bed. This was some five hundred yards long, but was planned to extend to two miles. Out beyond it an 'island', to which the jetty would ultimately be joined on completion of the western breakwater, was built from heaps of stone. But, as had now been proved, there was almost no stone in the Isthmus, none nearer than the Attaka quarries at Suez, which were still inaccessible. Until later, when a form of concrete was made on the spot, from imported lime and local sand, all the stone for the new harbour had to be brought from the quarries of Mex, beyond Alexandria, a journey of 150 miles by sea.

The initial problems of the construction of Port Said were thus problems less of engineering than of transport—in essence, problems of organization; and the same applied at this stage to the construction of the Canal as a whole. Uppermost in the minds of the Company's team, throughout these first three years, were the difficulties of supplying builders with materials, workshops with tools and machinery—above all workers with food and water, a commodity as scarce as stone.

These were years when the choice lay literally between life and death, not merely for the scheme as a whole but for the men engaged in it, tens of thousands of whom would be labouring in a waterless desert. In search of water, wells had been sunk from one end of the Isthmus to the other. But with a few exceptions the water found was so saline and bitter as to be undrinkable by man, and often even by camels. It was to meet this problem that the Sweet Water Canal was designed, and the *rigole*, the service channel, which between them would supply water along the whole line of construction.

But several hard and thirsty years must elapse before the completion of either. The Sweet Water Canal was at last to be finished in February 1862, bringing fresh water to the heart of the Isthmus for distribution to the stations around. Later that year a pumping station was established at its junction with the lake, which conveyed water by pipeline along the salt water stretch of the *rigole* to Port Said; and at the end of 1863 the southern, fresh water half of the *rigole* was to reach Suez. Meanwhile, since men must work and the Canal must grow, machines were introduced for the condensation of steam into water, which was often of inferior quality. But still the greater part of it had to come overland, for long distances and at great cost, from the Nile valley by fleets of boats across Lake Menzaleh, and by daily caravans across the desert, carried by thousands of camels.

*   *   *

But a crucial problem still confronted the Company—that of the supply of the labour itself. It was now Britain's firm intention, agreed between Bulwer and Lord John Russell, to arrest the Canal works by a campaign, in alliance with the Porte, against the use of forced labour in the Isthmus. Officially, the first indication of this policy was given in the House of Commons on June 25, 1861. To a question by Mr. Darby Griffith as to whether the practice accorded with the various 'humane edicts of the Ottoman Empire', Russell replied that the British Government had objected to it at the Porte, and was now investigating the position in Egypt itself. Meanwhile however the Viceroy, who was obliged under the terms of the Concession to supply the Company with labour, remained reasonably consistent in evading such pressure.

At first, in anticipation of it, attempts were made to recruit free labour, not merely in Egypt itself but in Syria and Palestine. The Philistines of Gaza seemed especially willing, and Lesseps himself led a recruiting campaign on a journey from Kantara to Jerusalem, where he was gratefully remembered by the Palestinian Arabs for pleading their cause with Mohammed Ali in his Consular days. But the response proved insufficient as the works progressed, and from 1861 onwards the Company was calling upon the Viceroy for *corvées* on a mounting scale. Thanks to these, the number of Egyptian labourers in the Canal zone had increased by the end of 1862 from 2,500 to 25,000 with promises of further contingents to come.

In the winter of 1861 Lesseps, with a shrewd sense of timing, conducted the Viceroy on a tour of the centre of the Isthmus, which disposed him towards further requisitions of labour. He was happy to sail down the Sweet Water Canal, gave orders to his Governors for the labour required for its completion to Timsah, and started to talk of its continuation, independently of the service channel, as far as Suez—a work which should serve Lesseps as a pretext for an extra levy of 50,000 men. The Viceroy made a triumphal entry into Toussoum, in a carriage drawn by six mules and preceded by a cavalcade of richly caparisoned dromedaries. His stalwart Viceregal Guard, 250 strong and mounted also on dromedaries, preceded him as he drove through two long ranks of labourers. Adding their voices to the strains of an Egyptian military band, they cried out 'Long live Said!' Before returning to Cairo he was presented with specimens of the various plants grown on the land which it had proved possible to irrigate—a load of maize, cabbages, cauliflowers, water-melons, radishes, potatoes, grasses, and a variety of ingredients for salads.

Even Colquhoun, who deplored the Viceroy's 'indiscretion' in visiting the Isthmus, and thus openly identifying himself with the Canal, had to admit in his report to London that he was received 'with every possible demonstration of joy'. Back in Cairo, Said gave a birthday party to which Lesseps was invited. Talking of his journey, the Viceroy declared that there should be no more 'anti-Canalists' in Egypt, and that he himself was now the real President of the Company. Turning to an English guest, he invited him to accompany him in the spring of the following year on a journey by boat from Port Said to Timsah.

In fact, he was a mere seven months out in the date of his invitation. Early in the following year, the Sweet Water Canal was completed as far as the Lake. But the sea water *rigole* from Port Said took longer, on account of the obstacle placed in its way by the plateau of el Guiser. This was a huge dune of sand, some thirty feet high and ten miles broad, constantly shifting and thus tending to fall in wherever a space was dug out. To maintain a cutting through it involved the ultimate excavation of 50 million cubic yards of sand, a task which was to take almost as long to complete as the whole of the rest of the Canal. But, thanks to the labour force now available, enough of it was removed by hand and a preliminary channel cut deep enough to take, by the end of that year, a stream of water, still shallow but just deep enough to permit the operations of dredgers.

On November 18, 1862, at eleven o'clock in the morning, the waters of the Mediterranean flowed into Lake Timsah. A special train, graciously provided by the Viceroy, transported Lesseps and his guests, who included the Consular representatives of the European countries, from Cairo to Zagazig. Next day they visited the 'Wadi lands', by the old Canal between Zagazig and the Nile, which the Company had bought from the Viceroy. Acquired for the dual purpose of supplying Nile water to the Sweet Water Canal, and irrigating new lands for the shareholders' profit, they were developing into a fine agricultural domain. Thence the party was conveyed down the Sweet Water Canal to the growing town on the shores of the lake, destined to become Ismailia but still named Timsah. On the landing-stage they were welcomed by the national anthems of France and of Britain— 'as it were', Lesseps recorded, 'an invitation to union, on the very ground lately believed to be a source of discord between the two countries'.

They then proceeded to the spot where the Canal, having at last penetrated the intractable plateau of el Guisr, was to join the lake. Here a triumphal arch had been erected in honour of Mohammed Said, at the end of an avenue of columns adorned with agricultural implements. This led to a Viceregal kiosk, built at his command and now surrounded by Venetian masts, from which fluttered bright-coloured pennants. Beyond, at the foot of a long low ridge, which was to become the Asiatic bank of the Canal, lay a trench, containing a stream of water fifteen metres wide—salt water from the Mediterranean, ready at the appointed moment to precipitate itself into Lake Timsah.

Deputizing for the Viceroy, Ismail Bey, his heir-apparent, presided over the men whose years of toil had achieved this consummation, together with the Grand Mufti of Egypt, the principal *ulemas* of Cairo, the Sheikh of Islam, and the Catholic Bishop of Egypt, with his clergy around him. Lesseps called for silence. Then, addressing the labourers massed on the dyke, which still dammed the waters, he proclaimed: 'In the name of His Highness Mohammed Said, I command that the waters of the Mediterranean be introduced into Lake Timsah, by the grace of God!'

All eyes were turned on the dyke. In a solemn silence the workers cut the dam. Then, with a cascade of water and an avalanche of soil, the Canal broke through into the lake. There was a roar of enthusiasm. The 'Bravos!' of the French vied with the 'Hurrahs!' of the English. Tears coursed down sun-bronzed cheeks as a band broke into the Egyptian national anthem; the *ulemas* called loudly upon Allah, and the Grand Mufti intoned a sonorous *fetva*, for repetition throughout the mosques of Egypt. It gave thanks to the Almighty for this grand and beneficent enterprise, and for this Frenchman of noble birth, the elect of God, who had accomplished His will with such high intelligence and uprightness of character. It sought His protection for the reign and the life of Mohammed Said, a ruler who worked day and night for the happiness, the prosperity and security of his people. Later, in the chapel at el Guisr itself, a *Te Deum* was sung by the Bishop and choir before a large and devout congregation. As Lesseps afterwards wrote of the occasion: 'What the ancient world could not do the modern world has done. Today, the union of the Mediterranean with the Red Sea must be regarded as an accomplished fact.'

*   *   *

Such also, to his own distinct surprise and discomfiture, was the conclusion of Sir Henry Bulwer, the British Ambassador in Constantinople, who visited Egypt some weeks later. Bulwer, after a visit to Upper Egypt 'for the sake of his health', made a tour of the northern half of the Canal zone with Lesseps. He lodged by the waters of Timsah in a wooden building, 'half hut, half house, such as is frequently seen in the outposts of American civilization'. He was impressed by the plans for its port, considering it 'difficult to conceive a better position for a large city', and noting that people were already beginning to flock to it. He sailed up a stretch of the maritime canal, still only some six feet deep and some fifty feet wide, but at Kantara saw its whole intended breadth, 'which certainly has an imposing appearance'. He inspected Port Said, which now had 1,500 European inhabitants, and made a detailed sketch of the plans for its harbour.

He found Lesseps—always a personal friend if a political enemy —'a most active and intelligent man in constant good humour and indefatigable'. He was struck by the energy and skill of the Frenchmen around him, stimulated by their greatest national sentiment, *amour propre*. Throughout his tour, he obtained a very different picture from that prevailing at home where, as he expressed it, there was a tendency to underrate the extent of the work done and to overrate the difficulties of the work still to be done. Now that the preparatory phase was over, this work was 'being pursued with an energy and an *ensemble* which is striking. . . . The only question as to its completion is a question of money.' What Palmerston had pronounced an impossibility, a bubble scheme, was now to become, in the hands of France and to the exclusion of Britain, a reality. 'I have seen all this', Bulwer reported to Lord Russell (as Lord John had now become), 'and till I saw it, I confess I did not conceive it.' In conversation with

Lange, Lesseps's agent from London he went further, admitting errors in his master's policy and remarking: 'What a pity we have no part in this affair!'

All this the Ambassador saw; and at the same time he was quick to see its dangers. Fears born in his mind as abstractions were now assuming around him all too concrete form. Palpable to his eyes was the metamorphosis of the deserts of the Isthmus into a flourishing outpost of France. Here, at Port Said, Timsah and Suez, were growing French towns, French lands around them and, along the banks of the Sweet Water Canal, entering into cultivation to become French territory, an enclave of Egypt already governed in practice by French authority. At Port Said, for example, there was 'a sort of Egyptian Governor, but the whole authority rests in reality with the Company'. The Egyptian Government would 'become a cipher, as well as the Porte'. Such was the moral of Britain's boycott of the Canal and refusal to invest in it.

Nor was this all. The French were employing monthly levies of 20,000 Egyptian labourers from various parts of Egypt. Calculated on the basis of 20,000 at work, 20,000 on the way, and 20,000 on the way home, this meant that 'nearly 60,000 men a month are taken from the agriculture of the country'. This, as Bulwer saw it, would 'not merely extend French influence and the name of France over the places bordering on the different canals; the whole population up to the Cataracts will have been gradually brought under the habit of being in the employment and actually under the Government of the French. It is impossible not to see the serious importance of this state of things, particularly when taken in conjunction with the naval power now organizing by France in the Red Sea.' Here was a potent political argument for reinforcing, on the 'general grounds of progress and humanity', Britain's objections to the French use of forced labour.

Next, Bulwer investigated in some detail the Company's financial position, and reported unfavourably upon it to London. Now that the works on the Canal had reached this stage, the problem of finishing it was primarily a problem of finance. So

Bulwer rightly judged. According to his calculations, on the basis of what had been done and remained to be done, the Company had started with a sum equivalent to £8 million, in hand or on call from the Viceroy. But the completion of the whole Canal—including the harbour of Port Said—would require nearly double that sum. He was inclined to believe that, by some means or other—further contributions from shareholders in view of the capital assets of the lands and near-completed works, intervention by the French and other Governments on general grounds—it would none the less be completed. Given the character of Mohammed Said and the unlikelihood of positive action by the Porte without positive support from Britain, 'the Company will, if pushed to the wall, get through these difficulties if the Emperor Napoleon and Said Pasha live and reign in their respective countries'

The question remained: how far would it matter to Britain if the Canal were completed or if it stopped short of completion? Bulwer doubted its commercial value, on the grounds that 'the time of canals seems to me pretty well past, since that of railroads commenced', and reverted to the hoary British argument, favouring the improvement of the Egyptian railways and thus the diminution of the Canal's value. On the other hand he doubted its strategic dangers, arguing belatedly, as his Prime Minister's opponents had previously done to deaf ears, that 'If we still have the upper hand at sea, we should under critical circumstances be more likely to get hold of [the Canal]. With a fleet at Port Said, and a small military force, coupled with naval means, at our disposal, we should then surely have the mastery over a most important part of Egypt, as well as over the communications with Syria under our power.'

Thus the Canal, if completed, would not be a danger to British interests; nor would failure to complete it be a loss to British interests. What affected these interests most was the high degree of influence which the Canal Company, through its lands and its towns and its hold over the Viceroy, threatened in either case to wield over Egypt. Bulwer, in proposing to London a course to pursue, no longer sought to prevent the Canal from being made.

He sought to prevent it from being made by the men at present making it, and on the terms and with the perquisites they at present enjoyed. According to his view, it was 'not so much the Canal we have to combat, as the French Company and the power it is becoming possessed of'.

The enemy, in short, was now the Suez Canal Company. The British Government should watch carefully its financial affairs, and should be ready, if and when breakdown threatened, with a proposed solution, agreed with the Porte. Here the crux was the matter of the Viceroy's payments, the bulk of which were not yet made. Lesseps was still urging him to take up a loan for their immediate discharge, in place of their discharge by instalments.

Instead, so Bulwer in effect suggested, the Porte itself, with the backing of Britain, should propose a loan, designed to assist not the continuation of the Company but its disappearance from the scene, its capital reimbursed with a fair profit for the works carried out. The Canal would then be sanctioned by a firman of the Sultan, and the whole enterprise transferred to the Government of Egypt and the Porte itself. As Bulwer wrote privately to Lord Russell, one of two things might happen. Either the Company would break down before the Viceroy's payments were required; or, 'when those payments have to be made the Porte may interfere in some effectual manner in the affair. This however it will not do except under the influence of a clear promise of protection from us.'[1] Such were the ideas now simmering in the mind of the successor of Lord Stratford de Redcliffe.

Meanwhile he had twice seen and talked at some length with the Viceroy. That summer Said had paid a visit to London, where the Canal was not officially mentioned, but where he was made aware of the prevalent coolness towards it. This had induced him to call, on his own account, for a report on its prospects from a leading British engineer, Sir John Hawkshaw, 'well knowing such advice would be loyally given, irrespective of party feelings'. The report, more favourable than any of its precursors, vindicated the work in progress and encouraged him greatly. But now he

[1] Bulwer Papers: December 16, 1862.

was in a mood of despondency, self-pity and self-reproach, which Bulwer sought to turn to diplomatic account.

Said was deeply concerned as to the state of his health, and went so far as to remark: 'I am a finished man. I shall never recover from this illness, and if I do so, I shall retire from public affairs.' He blamed himself for his financial errors, but as soon as he had remedied them, 'I shall relinquish power, if I am still alive—and then come what may. I am demoralized.' Talking of the Canal and its weight on his shoulders, he said again: 'I am discouraged, I am ill, I am lost.' On an official visit to Paris, where he had expected more support than in Britain, he had been depressed by the apparent lack of enthusiasm for the Canal, in official circles. He seemed, if only for Bulwer's benefit, disillusioned with the French. They had 'made big eyes' at him, but what else? Now the Company was pressing him to negotiate this loan, to pay for all his shares. But he had agreed to nothing.

Here was Bulwer's opportunity. Listening patiently to his troubles, giving him sage, sympathetic advice, he won his attention and apparently also his confidence. Enlarging both on the financial perils confronting the Canal, and on the political perils of French influence confronting Egypt through the Company's control over labour and land, he drew from the Viceroy two promises: not to release more labour beyond the present levies; not to agree to any loan or issue of Treasury bonds beyond his original financial agreement. If the Viceroy held to these promises, the British Ambassador promised in return his own friendship, and he would do his best 'to get him out of his embarrassments without any peril or loss of honour'—and without annoyance even from the French. 'A great weight', commented Bulwer, 'seemed to be taken off his mind when I spoke to him in this manner.' But he felt bound to add that he might not act up to his words.

Mohammed Said was in fact a genuinely sick man, suffering from a disorder of the stomach and the intestines. Soon after Bulwer had left, Lesseps visited him at the Nile Barrage and found him in pain, in the hands of his doctor. He sought to console him with the reassurance that it was no more than an attack of sciatica, which would have passed by the morning. Next day he was better, and they sat talking easily, in the garden. Familiar with his master's nervous temperament, he liked to hope, despite the doctor's discouraging reports, that his present state might be due to the strain and fatigue of Sir Henry Bulwer's visit. But when a personal servant appeared, carrying a pipe for Lesseps, the Viceroy remarked in a low voice: 'You see that fellow there? He is watching me attentively, because it is he who will announce my death to my successor, and according to custom he will get a good tip.' Since however after two more days Said still seemed better, Lesseps left, as he had planned, for a tour of the Isthmus.

Ten days later, as he was sleeping in the desert on his way north, on horseback, towards Kantara, he was awoken in the middle of the night by a messenger, with a telegram. It announced that the Viceroy had been moved to Alexandria, critically ill, that he was now *in extremis,* and that if Lesseps wished to see him again he should come at once. He rode through the night to Ismailia, thence took a boat, drawn by two dromedaries, up the Sweet Water Canal. Another boat met him at Tel-el-Kebir. On board was Voisin Bey, now his Director of Works, who told him that the Viceroy had died that morning.

This was despairing news for Lesseps—not simply, as he afterwards wrote in his Journal, 'because of my enterprise, in which my faith is entirely serene, come what may; but because of this cruel separation from a faithful friend who, for twenty-five years gave me so many proofs of his affection and confidence'. Arriving in Alexandria after a twenty-hour journey, he was admitted to the mosque where the body of Said was lying. For an hour he remained there, quite alone with his devotions and memories, his head bowed and resting on the turban of his dead pupil and master.

CHAPTER XI

Ismail takes over

Ismail Pasha, who now became Viceroy of Egypt at the age of thirty-three, was very different in character from Said. He had none of his uncle's impulsive, confiding ways. As dry and secretive as Said was warm and expansive, he was slower but craftier in his mind, given to orderly habits and a methodical attention to business. Comparing the two successive rulers, the French Consul-General, Beauval, observed that Ismail lacked the charm and the generosity of Said, and his indefinable touch of the grand manner. He was less of a Prince than his uncle but more of an administrator, an orderly worker with a more serious mind and a more far-sighted outlook. With his industry and intelligence, he had many good schemes for his country, but was apt to be busy, as Bulwer observed, with too many at once—'a multiplicity of affairs creating confusion'.

Ismail, short and still slight in figure, though later turning to corpulence, was unattractive in aspect, with curiously large and ill-shaped ears. But for what he lacked in personal charm he compensated in polish and courtesy and a seductive manner, which won from those in his presence at least a momentary feeling of confidence. For Ismail was a shrewd judge of human character. While talking he would observe a visitor with one eye half-closed and another wide open, provoking the comment that he heard with one eye and spoke with the other—a comment which he once amplified to a friend with the remark, 'Yes, and I think with both.' He knew how to please, when he chose to do so, and could be all things to all men, remarking once: 'One man rides a horse, another a donkey, another a camel, which have all different movements. The best rider is the man who can ride all three equally well.'

Ismail was no man of dreams, as Said had been, but a man whose governing force was ambition. A hard-headed realist, he sought power for its own sake and believed that its mainspring was money. Self-interest, he cynically believed, governed mankind: every man had his price. In his own attitude to money there were sharp contradictions. As one of Egypt's largest and richest landowners, he had built up and administered his estates with a peasant's tenacity and thrift, taking care of his pence and his crops with a strict attention to practical detail. Said compared him contemptuously to a trader, bent on paltry gains.

But he was to take little care of his pounds when his patrimony came to him. In Ismail's nature there was much of the speculator, side by side with the peasant. Surrounded as he was by men piling up fortunes, usually at his own expense, he envied and sought to emulate them. Visiting the Bourse in Paris, he once exclaimed: 'If I wasn't Viceroy, I should like to be a stockbroker.' Throughout his reign he squandered millions, not merely on reckless speculations but, in the grand manner, on hospitality, on luxury, on pleasure. He spent with a prodigality which far outdid the wilder extravagances of any of his forbears—and which was to prove in the end his undoing.

Withal, in Ismail's nature there was an inherent timidity and caution, a suspicion of the motives of others and a fear of incurring their enmity. As to the Suez Canal, he had still to make up his mind on the best course to pursue in his own interests. He approached the problem in a spirit of wary appraisal of the contending forces and the issues confronting him.

Seeking, in an equivocal style, to please everyone, he started by pleasing the French. To Beauval, on the day after his accession, he declared: 'I am more of a Canalist than Monsieur de Lesseps, but in a positive spirit. I believe that no work is so great or will be so productive for Egypt. But at present its foundations are uncertain and ill-defined. I shall affirm them and then, far surpassing my predecessor, shall press the works to their completion.' He pleased the British by repeating, at a meeting of the Consular Corps, that the system of the *corvée* would definitely be abolished throughout the country. But, in response to a French enquiry,

Ismail Pasha, Khedive of Egypt

The Emperor Napoleon III and the Empress Eugénie

he added that he was referring to the country as a whole and not to the zone of the Suez Canal in particular. In fact, after a brief delay, the monthly contingents were still despatched to the Isthmus.

Ismail then displeased the French and pleased the British by proceeding without further delay to Constantinople to be invested, on his accession, by the Sultan, and to obtain instructions on this and other matters concerning the Canal. When Beauval protested that these instructions would come from Britain and were therefore all too predictable, the Viceroy replied: 'The Porte will not reply definitely. . . . If the reply is too much delayed, I shall not wait for it; if it is equivocal, I shall act regardless of it. Besides, between ourselves, my investment alone will make me morally free in my actions.'

The British Ambassador now saw his chance. Bulwer had talked with Ismail in Egypt before his accession, finding him opposed to the Canal and in particular, as a landed proprietor, to the transference of labour to the Isthmus. He had since pressed upon London his own previous view that the Company should be wound up, and he now urged that the moment for this had arrived. But he went too far in suggesting that the Canal should be carried on by Egypt, under a European guarantee. To this Lord Russell replied firmly that 'the British Government can in no case guarantee, promote or favour the Suez Canal, which they would wish to see abandoned'. He proposed, nevertheless, that the Sultan should come to a compromise about the shares taken by Said, and 'get off as cheaply as he can.' In conjunction with the Viceroy he should prohibit forced labour, and the Canal should then be left to private speculation, 'without Government favour'.

To Bulwer's momentary discomfiture, Ismail the Viceroy talked in a different tone from Ismail the Heir-Apparent. In their first interview he said to the Ambassador: 'Certainly the Suez Canal is a great misfortune for Egypt and for myself. But we may often want to check an evil without being able to do so.' He foresaw difficulties in suppressing the *corvée* and thus provoking the importation of French in place of native labour. The whole question, he concluded, was very difficult and 'I am not strong enough to stand alone against France'.

Bulwer, countering Ismail's line, which he took to be inspired by the French, brought up the predictable British arguments with force. Spicing them with implied threats, he stressed that Britain, had she chosen, could have taken over Egypt long since—and that quite recently Russia had urged her to do so. She had preferred to keep her free of foreign influence by leaving her in the hands of the Viceroy and of the Porte. But if they 'were inadequate to fulfil their duty and maintain their position, if they were merely a cloak for foreign usurpation, the sooner they were swept away the better'. He enlarged upon the evils of the *corvée* and the perils of the grants of land to the company which would make Egypt a French province.

With the French Ambassador, Ismail followed a different line, declaring: 'I am persuaded that France alone has a policy entirely sympathetic to Egypt. I shall devote all my energies to the realization of the work of the Canal.' He was at pains to relieve the Ambassador of any anxiety as to its future. The Canal would be of incomparable value to Egypt, its construction would not be interrupted, and he was determined to complete it, removing all the obstacles placed in its way by Said.

With the Grand Vizier and his Ministers, Ismail sought to evade discussion of the Canal, taking the line that it was up to the Porte to state its own opinion, for which he had asked, and to give him instructions if need be. But let it do so promptly; local action would be required of him within two or three weeks. He refused to attend meetings of the Council of Ministers, or to accept invitations to dinner. Entertaining the Ministers to dinner himself, on the eve of his departure, he evaded discussion on the subject. Finally, paying an official farewell call on Aali Pasha, the Foreign Minister, he found Bulwer also present. Together they urged him that the present state of affairs should not be allowed to continue. Thus the Viceroy agreed, before his return to Egypt, to leave a note on the Canal for the Porte.

This would reply to questions (optimistically drafted by Bulwer before Ismail's arrival) concerning the Viceroy's debt to the Company; the utility of the Canal to trade in general and Egypt in particular; the number of workers at present transferred from the land to the Canal works, their rates of pay, and the effects on agriculture and the country's general welfare; the political effects on both Turkey and Egypt of the piercing of the Isthmus and the possession of lands by the Company; the possibility of making the Canal under other conditions—for example by the Egyptian Government. The Viceroy was invited to give his views on these questions, and on the solution of the Canal problem in general. His reply would serve as a declaration of his attitude towards it, and as the basis for a decision by the Porte, for communication to Paris and London.

But Ismail had no intention of thus committing himself to a statement of policy. His reply, delivered after his departure, was brief and evasive. His debt to the Company was a debt of State; the commercial and other value of the Canal was open to question; the cost of it was beyond his capacity to judge; the workers furnished to the Company amounted to 20,000. With regard to the other questions, the Viceroy would, on his return to Egypt, obtain the relevant documents and despatch them to the Grand Vizier. The Porte was stunned at such repayment by the Viceroy in its own negative coin. Sir Henry Bulwer was himself taken aback, but at once embarked on an alternative and positive course.

*   *   *

Ismail's diplomatic reconnaissance in Constantinople had helped him to reach some significant conclusions. He now had a clearer appreciation of his position, and of the immediate course he should pursue. From the Porte, so his journey had proved, he could expect nothing. It denied him power, yet refused to take the responsibility that its own power imposed upon it. Britain too denied him power. The Turkish and British Governments were clearly combining to diminish his authority in the internal affairs of Egypt—in this case contesting his right to dispose of his own labour and lands as he thought fit for his country's economic development.

The Suez Canal, which would increase tenfold the political importance of Egypt, had grown into a symbol of her increased independence. Britain, by keeping her in subjection to the Porte, was determined to curb this independence, hence laboured still to stop the Canal's construction. Ismail wanted it, and thus now supported the Canal, as a source of increased power, in terms of a hereditary dynasty, with the right to pass on the succession to his eldest son, and of autonomy in financial and foreign policy. Mohammed Said, too, had coveted this, but to Ismail, now that the Canal, which could be its instrument, was no longer a dream but a probable *fait accompli*, it was a deep-rooted ambition —more particularly because his two potential heirs-apparent were his bitterest enemies. Neither Britain nor the Porte would help him towards the power he craved. Only France was pursuing a policy sympathetic to Egypt, hence possibly to Egypt's increased independence. Clearly France was the ally who would best suit his interests.

On landing in Alexandria, the Viceroy gave an especially warm welcome to Lesseps, paying him the compliment: 'Had you been Viceroy of Egypt, as well as President of your Company, you could not have looked after the business of the Suez Canal better.' Lesseps had lost no time in officially naming Timsah Ismailia 'in honour of the Prince who will continue and complete our great enterprise, and will at the same time bring back to life for the Arabs, called to fertilize the waste lands of the Isthmus, the memory of the founder of their race, the son of Abraham and Hagar'. Ten days later the Viceroy, swiftly forestalling the next move from Constantinople, agreed with Lesseps on two Conventions. They were drafted, drawn up and signed with a stroke of the pen in a single afternoon. Their signature marked a new step in the French Company's progress.

The first allowed for the construction, no longer by the Company but by the Egyptian Government, of the Sweet Water Canal from Ras-el-Wadi to Cairo. This reach, covering half the channel to the centre of the Isthmus, was designed to by-pass the Damietta branch of the Nile, and to feed the Canal from the river's main stream. In return for this, the Company relinquished all claims, stipulated in its Concession, to the lands on its banks which would thus be brought into cultivation.

The second, and to the Company the more important Convention, was financial. It settled at last the long-delayed payment for the Viceroy's shares. Accepting the obligations of Mohammed Said, Ismail agreed to pay at once the balance owing for 177,642 shares, on the first call of 100 francs per share—a total of 15 million francs. On the second call of 200 francs per share, already paid up by the other shareholders, he agreed to pay three-fifths of the sum due, in monthly instalments, starting from January 1, 1864. This represented a further total of 35 million francs, payable in Treasury bonds at the rate of a million and a half francs per month. Thus by 1866—in three years' time—Ismail should have contributed to the Company a capital sum amounting to 50 million francs (£2 million).

The payment of the remaining two-fifths, then likely to be due on the final call, in terms of the full value of 500 francs per share, would be negotiated between the Company and the Egyptian Treasury, in the light of its circumstances at that time. On Ismail's present showing, the discharge of this, bringing the Egyptian Government's total investment to some 88 million francs, should not be in reasonable doubt. At a time when the Company's cash in hand amounted to 12 million francs and its expenditure to nearly 2 million francs per month, this represented a major relief from the financial anxiety which had weighed heavily on Lesseps's shoulders for four years past.

The two Conventions, and the rapidity of their negotiation and signature, was a political *coup* for Ismail, which caused dismay in Constantinople. It was, as Bulwer reported irritably to Russell, 'a stupid and treacherous measure'. He laid the blame for it on Colquhoun, his easy-going Scots Consul-General, whom he had been briefing assiduously on 'the necessity of spirit' in counteracting the influence of the arrogant French on the Viceroy, but who was clearly no match for Ismail's wiles.

As Bulwer wrote: 'Whilst I have been settling Canal business here and carried my point, Colquhoun has allowed the Viceroy to make new Conventions with Lesseps. It is too bad. . . . After this I can answer for nothing.' A strong message from His Lordship should be sent to Colquhoun. Already Palmerston (now in his seventy-ninth year) had been grumbling about the Consul-General: 'Mr. Colquhoun ought certainly to be stirred up with a long pole to rouse himself to act with energy in support of the Viceroy. I am afraid he is wanting in this respect. . . . When I saw Mr. Colquhoun before he returned to Egypt, he spoke very favourably of Lesseps saying that he is a perfectly honest and honourable man. Now as Lesseps is to speak in plain terms a rogue and a swindler this opinion expressed by Mr. Colquhoun seems to imply that somehow or other Lesseps has the length of his foot and has trammelled his mind. . . . His not hoisting his Flag looks like a wish to keep well with the French rather than to support the Viceroy.'

To Bulwer's rebuke Colquhoun replied, with distinct spirit: 'Ismail Pasha deceived you all at Constantinople. Is it so strange he should deceive your poor subordinates here?'[1] He was now sharply instructed by Russell to use all his influence to induce the Viceroy to cancel the Conventions, which he had no right to make. A later instruction, drafted by Palmerston himself, pressed upon him the need to warn the Viceroy that, by disobeying the Sultan, he might put his tenure of the Viceroyalty in danger, and that 'if he continues to yield to Lesseps, he will soon be only the nominal while Lesseps will be the real Viceroy of Egypt'. Colquhoun was further enjoined: 'You are not to content yourself with being simply the mouthpiece of Her Majesty's Government in conveying such messages to the Pasha, but you should avail yourself of the weight which attaches to your high character . . . actively to enforce the counsels and opinions of the Government.' But none of this carried weight with Ismail.

Nor did the official visit of the Sultan himself to Egypt, at this time, further the cause of the British, who had strongly opposed it. Abdul Aziz was the first Sultan to visit the country since its conquest by his forbear, Selim I, 350 years earlier. He came no longer in the spirit of a conqueror, nor even, according to protocol, as his Viceroy's host, but as his guest, a distinction of status which he made publicly clear from the start. The splendours of the visit, of which a highlight was the Sultan's first journey in a

[1] Bulwer Papers: April 14, 1863.

train, enhanced Ismail's prestige rather than otherwise and, despite the intrigues of Bulwer's agents, did nothing to diminish that of his now avowed allies, the French. It also gave Ismail the chance to seek friends for his own personal cause at the Porte, by the lavish distribution of presents to his sovereign's Ministers and entourage.

*　　　*　　　*

Meanwhile, however, a counter-attack had been launched from Constantinople. Since the Viceroy, in the course of his visit, had refused to be drawn, then the Porte must at long last bestir itself to take the initiative. Within a week of Ismail's departure and the delivery of his non-committal reply to the Porte, Bulwer had persuaded its Ministers to draft a Note to its Ambassadors in Paris and London, together with another to the Viceroy in reply to his previous request for its opinion on the Canal. Three years earlier, after the Emperor's intervention, the Porte had thus approached the two great maritime powers for their views on the subject, but in generalized terms and with nebulous effect. So the question had lapsed. But this time its approach was specific and positive. Here in effect was the Porte's first clear definition of policy on the subject of the Suez Canal since it was first broached by the Viceroy of Egypt six years earlier.

As finally drafted, after insistent but largely ineffective efforts by the French Ambassador to achieve modifications of form and content, the joint Note to Paris and London was despatched on April 6, 1863. It made the significant admission that the Sublime Porte had no wish to prevent the realization of an enterprise which could be of general utility, making it clear that such a Canal, traversing the province of Egypt, would be subject to the Sultan's sovereign authority. The Porte's consent to the Canal was contingent on specified conditions, safeguarding on the one hand its external security and on the other the internal interests of Egypt which the Sovereign was obliged to protect.

First, there must be international guarantees of the Canal's neutrality, on the basis of those granted to the Dardanelles and the Bosporus. Secondly, in the course of its construction, forced labour must be abolished, as in all the Sultan's dominions. Thirdly, the Company must abandon lands conceded to it, in the neighbourhood of the Sweet Water Canal. If these conditions were agreed, the Porte would be willing to discuss with the Viceroy, in co-operation with its 'two most sincere allies', the other clauses of the Viceroy's Concession. If, under these conditions, the Company preferred not to continue the work of construction, the Porte would seek means of reimbursing it for the expenses it had already incurred.

The Note from the Grand Vizier, despatched at the same time, was couched in the same terms, but more strongly worded:

> Neither the Sublime Porte nor the Powers, keenly interested in the progress of civilization in the East, can allow this great work to be accomplished by a system of labour severely and for ever condemned by all civilized nations; a system of which a disastrous consequence has been to remove sixty thousand Arabs from agriculture, industry and the trade of the country. Neither can the Sublime Porte recognize that further stipulation which concedes to a foreign company a considerable portion of Egyptian territory. All that the Government of his Imperial Majesty can do . . . despite the lack of respect which the Company has hitherto shown to the rights of the territorial Sovereign, is to grant it preference, if it will now eliminate from its programme the items we do no accept. In this case . . . we shall examine in good faith and without prejudice all the other conditions of its contract. If the Company does not wish to proceed with the enterprise, with the acceptance of these stipulations, the Sublime Porte is ready to discuss with Your Highness the adoption of means to indemnify the Company for its expenses and to devise a more suitable and practicable plan of execution.

Meanwhile the Viceroy was instructed to withhold all forced labour and to postpone work on the extension of the Sweet Water Canal to Suez. Thus the preliminary skirmishes were over, and war was finally launched by the Porte against the Canal Company, bringing into full play the two formidable weapons—those of the labour and the lands—which Britain had been holding in reserve for this purpose. In London *The Spectator* exulted: 'The Suez Canal, which was to have turned the current of Asiatic trade and poured the wealth of the richer part of the world into the lap of Marseilles, has been blocked up by the Sultan, and France, woken up from a brilliant dream of ships, and commerce, and empire, rages against the "perfidious" power which so jealously guards all three.'[1]

*   *   *

The question of forced labour had been raised once more in the House of Commons by Mr. Darby Griffith on May 16, 1862, in the interests of humanity and in view of the fact Britain 'had always evinced great interest in the prevention of slavery in all parts of the world'. A great evil, he asserted, was being perpetrated by the Suez Canal Company 'in an unblushing manner'.

This debate provoked Lesseps to write at length to Henry Layard, the Under-Secretary of State for Foreign Affairs, seeking to put the facts in their proper perspective. First he raised the question as to whether one Government had the right thus to interfere in the internal affairs of another. The principle of slavery was accepted in America and until lately in Russia. Had Britain, in defence of the rights of man, protested against it in Washington or St. Petersburg? He quoted Dickens to show that in Britain itself apprentices were the saleable property of their masters from childhood until the age of twenty-one, moreover subject to corporal punishment and deprivations of food; while Mr. Layard, he knew, was familiar with the plight of the labourer in the British provinces of India, and had zealously championed his cause. But would Britain not be indignant if France in the name of humanity, were to protest against his condition and that of the British industrial apprentices?

As to conditions in Egypt, Britain could hardly deny the right of the Egyptian Government to recruit labour—as in all oriental

[1] May 16, 1863.

countries, for public works of general utility. She had invoked it
herself, from the Viceroy Abbas Pasha, for the construction of the
railway from Alexandria to Cairo and later for its extension to
Suez. These rails, as Lesseps was wont to express it, had been
laid on 'thousands of Egyptians corpses'. Only a few months
ago a labour force, fifty thousand strong, had been similarly
raised to repair damage to the railway from storms and Nile
floods, and thus ensure rapid resumption of the mail service
between Britain and India. Nor at any time did the workers on the
railway (or, he was afterwards to claim, the present workers for the
P. & O. Company at Suez) enjoy the advantages now accorded to
those on the Canal.

Here labour conditions, contracted between the Canal Company
and the Viceroy, were in the general interests of the Egyptian
population and of humanity as a whole. The Canal workers were
given payment above the normal rates; they were assured good
nourishment; they were not subject to corporal punishment; they
were given free medical services and, when sick, continued to
draw half their pay. These were conditions generally applauded
by Europe on their first introduction in 1856.

But now, six years later, the House of Commons, inspired
largely by misapprehensions, had elected to criticize them.
Lesseps disposed of its criticisms one by one. The workers were
paid directly, in cash on the spot, not indirectly through their
sheikhs or in notes negotiable elsewhere, as speakers in the
House had declared. Nor were they ever disbanded before being
fully paid. The mortality rate among them amounted hitherto to
two men in ten thousand. Their combined wages added up to
millions of francs, which would go to improve their lands and
to release them from the hands of the moneylenders. The fellah
of Egypt was thus being progressively raised to the status of a
free worker.

In fact the fellah himself, at this stage in his evolution, could
not always be expected to see his prospects in so rosy a light.
Reluctant to work at all, except on his own plot of land, resentful
at being torn from it, perhaps at harvest or sowing time, to work
in a remote and unfamiliar part of the country, he was liable to

desert when he had money in his pocket, so that the Company took to paying him only at the end of his monthly stint.

Undoubtedly anomalies and abuses occurred. Bulwer, after his visit to the Isthmus in the winter of 1862, maintained that the workers were in fact paid not more but less than the normal rates and that, though the Government brought them from their villages, they were obliged to find their own way home (an objection applying, in principle, only to those not requiring rail or river transport). Nevertheless, his own real objection to the forced labour system—still prevalent in practice if not in theory in other parts of the Ottoman Empire—was less to the conditions of work themselves than to the ascendancy which it threatened to give to the French, at the expense of the Egyptian Government, over the working population. The danger, as he saw it, lay in the fact that the Company was treating its labour not badly but all too well.

The merits of the system for the general good, evident in practice and arguable in theory in a country so underdeveloped as Egypt, were not seriously contested by any Consular official, French or British. Colquhoun himself, on a journey with Beauval to the Isthmus, addressed a gathering of the Company's employees (so his French colleague reported) as 'veritable pioneers of civilization', and complimented them on the manner in which they were instructing this native people, 'so gentle and good, but still backward in civilization and industry'. Expressing his views to London, he wrote: 'I fully believe that, with time, when the Arab labourer finds he will be regularly and honestly paid for his labour . . . he will not look on an enterprise such as this in the light of a Government *corvée*, where he barely had the means of subsistence and was driven like a beast to his work.'

On the subject of the abolition of the *corvée* in general, which Ismail had foreshadowed soon after his accession, it was generally agreed among the men on the spot that the country could barely function without it. The annual upkeep of roads, railways, dykes, and especially canals, depended, in Colquhoun's words, on 'masses of men and lads brought at a fixed period to the work'. Masses were required because the fellaheen were poor and weak workers, eight or ten of them achieving less in a day than a single European.

There could thus be no serious question of paying them at European rates, no question either of putting the work on a voluntary basis if it were to get done at all, if the land were to be fertilized and the fellaheen themselves assured a modicum of food. The answer to the problem, in Colquhoun's view, was not abolition but a revision of the system and the reform of its abuses, the provision of wholesome food and tools, the limitation of hours of work, the establishment of proper restrictions. What Colquhoun was inclined to suggest to Ismail was 'the judicious application of the *corvée* for the public good'.

These were highly impolitic views for a British Consul to express at a moment when his Government was insisting on its total abolition at once, on high moral principles, and with the political motive of arresting the work of a powerful French Company. Colquhoun was instantly warned not to express them. In a long reply to his despatch, Lord Russell insisted that any temporary continuance of compulsory labour for irrigation channels must not become a precedent for the Canal or other speculative enterprise. Forced labour was 'a remnant of barbarous times and of a state of political and social ignorance'. If the fellaheen of Egypt were less effective workmen than Europeans, the inference was that they were paid less than Europeans. Their failure was due to 'a natural disinclination of all men in all parts of the world to exert themselves to the utmost in the performance of work for which they are not paid and which they are compelled by force and the fear of punishment to perform. . . . This forced labour system degrades and demoralizes the population and strikes at the root of the productive resources of the country.'

Such arguments could not reasonably be applied, in the view of the French, to the works of the Canal Company. For they were based on an abuse of the term 'forced labour'. Forced labour was unpaid labour—in short, a form of slavery. The labour in the Isthmus was paid—moreover paid at a fair scale of rates which enabled the worker to save. True, it was directed to the site by the Egyptian Government. But it was limited to a brief period and a fixed task for each worker. The Company's system was in fact designed to achieve the eventual abolition of that *corvée* which the English Liberal statesman deplored. It would be a civilizing influence, because it substituted work for that inertia which was Egypt's inherent malady; it would bring a young generation of Egyptians into contact with Europeans of industrious and orderly habits, opening their eyes to the achievements of European science, encouraging them to break free from the primitive methods of cultivation which had impeded their progress for thousands of years.

*   *   *

The second weapon conjured up by the British in support of the Porte's offensive against the Canal Company concerned the lands which it held by the banks of the Sweet Water Canal, now nearing completion to Suez. These, the Porte insisted, must be abandoned, since they involved alienation of Ottoman territory and the creation of foreign colonies incorporating several towns which would be almost independent of the Empire.

The grant of the lands, then mostly abandoned desert wastes, had been made by Mohammed Said in his original Concession, with a view to their reclamation by the Company. When Kiamil Pasha, in Constantinople, objected to the transaction, he replied pointedly with a quotation from the law of Islam, that he who made waste-lands productive was entitled to enjoy their use for as long as he paid his taxes. The fertilization of these particular waste-lands would not only prove a blessing to Egypt but might well also serve as a lesson to other provinces of the Ottoman Empire 'where bad administration and outdated prejudices have impoverished and depopulated the country'.

For the Canal Company the grant was an essential element in its economy, since the lands, when they came into cultivation, would provide its shareholders with a guaranteed source of revenue until the Canal itself came into operation and the payment of tolls could begin. They were legally entitled to this return on their investment. Moreover, from this the Egyptian Government would benefit equally, as owner of nearly half the shares.

The lands could hardly become foreign colonies, since the climate and general conditions were unsuited to European agricultural settlement, and those who leased and worked them would be almost exclusively Egyptians and other Ottoman subjects. Certainly Europeans—bankers and industrialists and businessmen—would inhabit the towns of the Canal Zone, as they already inhabited such cities as Alexandria and Cairo and, in other parts of the Empire, Beirut, Smyrna and Constantinople, encouraged to do so by the Porte, for its own benefit, and enjoying, under the laws of the Empire, the right to hold property and to establish commercial and other concerns. As elsewhere, this right gave them no local administrative or political power, since they would be subject, in their various districts, to the authority of the Egyptian Government. Lesseps scorned the Porte's fears that the Suez Canal Concession would place its Egypto-Syrian frontier under the rule of a foreign company. On the contrary: 'This frontier will continue to belong to Egypt. . . . But, instead of an arid desert frontier, Egypt will have a frontier which is fertile and populous.'

Meanwhile, to discuss ways and means of meeting this two-pronged offensive, Ismail Pasha sent an official representative to Constantinople. He sent his Foreign Secretary, Nubar Pasha, who was henceforward to wield an insidious influence over the affairs of the Suez Canal.

CHAPTER XII

Intrigues in Paris

Nubar Pasha was an Armenian Christian, not yet forty years old. Supposedly the cleverest man in Egypt, he was certainly one of the richest—the possessor of lands and a fortune skilfully amassed over a period of years by adroit and often devious methods. The archetype of the Levantine confidential adviser and middle-man, frequenting the ante-rooms of Ottoman power, he had been an *éminence grise* of the two previous Viceroys. Now, under Ismail, he had soared to a position of paramount influence—though not yet to that pinnacle of power from which he could jest that Ismail was 'the name with which I sign my decrees'.

Nubar was a man of restless ambitions, flexible principles and lively intelligence. Supple in mind, he was a master of words, whether in dialectic, in negotiation, or simply in the exercise of charm over those whom he sought to please or impress or manipulate. But for all his ruses, Nubar had a directness and frankness of speech which would disconcert those around him and would perhaps, for that very reason, rouse them, in an oriental world where euphemism and equivocation were prevailing conventions. Armed with a sense of essentials and an impatience of cumbersome detail, decisive in opinion where his master was hesitant, he would take the initiative when others were reluctant to do so. Lesseps, who had no great reason to love him, said of Nubar that he could be either very good or very bad: 'When he is good he is excellent, because he is very bold; he is the only Ottoman Minister who dares take things on his own shoulders.' All in all he represented, in the view of an Englishman who knew him well, 'what little of real statesmanship there exists in Egypt'.[1]

Nubar, after an education in Paris, was brought to Egypt as

---

[1] C. F. Moberly Bell, *Khedives and Pashas.*

a young man by his uncle, an Armenian in Mohammed Ali's service. It became one of his tasks to read aloud to the Pasha Thiers's history of the Revolution, Consulate and Empire—and thus to fire him, for better or for worse, with Napoleonic ambitions. But Nubar was to become (in the words of Beauval) France's 'most zealous, most constant, and most dangerous enemy'. It was to Britain that, from the reign of Abbas onwards, he turned his allegiance—to such an extent that he was generally believed to be in the pay of the Foreign Office. He looked 'to the Government of Egypt by England through an Armenian resident'. It was to the interests of Britain that he now lent his influence on arrival in Constantinople early in June 1863, to discuss the business of the Canal.

First he paid his respects to the current Grand Vizier, Fuad Pasha, handing him a reply to the Porte from the Viceroy, which bore only indirectly on the main points at issue, but also a gift from him of £30,000 which was gratefully accepted as a timely means of repaying a troublesome debt. Then he called upon Bulwer, who was happy to see him—respecting his ability, his authority with Ismail and his pro-British sentiments, and remembering the hostility towards the Canal which he had shared with his master before his accession.

Placing his cards on the table, Nubar declared that the Viceroy was ready to take back the lands of the Company, by purchase, and to pay for the completion of the Sweet Water Canal. On the other hand he was reluctant to abolish forced labour, since this would stop the construction of the Canal altogether, and antagonize the French Government. The concession of lands, he argued, was a danger, but the employment of forced labour was a mere inconvenience. It might be well to suffer the inconvenience in order to escape the danger. Bulwer warned Nubar that his Government would never agree to forced labour. But it was evident that the Company would be unlikely to agree to a sale of the lands if it were not to be allowed the labour; for the gain from one would be the loss on the other. Between them the two men worked on a compromise, covering both these points.

Nubar meanwhile called on the French Ambassador, the

Marquis de Moustier, to seek his support. He hoped that the French Government would put pressure on Lesseps to accept such a compromise. But Moustier, following his instructions from Paris, made it clear that his Government was concerned with the Canal question only in two respects. It called on the one hand for the completion of the maritime canal, and on the other for the protection of French capital engaged in the enterprise. The lands and the labour were matters of commercial detail, to be agreed between the Viceroy and the Company independently of official intervention. Nubar finally announced his intention of proceeding to Paris to settle them, and to this course Moustier indicated polite assent.

The compromise was agreed by the British Government and the Ministers of the Porte, for submission to the Viceroy in the form of a letter from the Grand Vizier. Under its terms the Viceroy would agree to negotiate purchase of the lands within a period of six months. He would declare the abolition of forced labour and introduce regulations for a system of free labour, at fixed rates, supervised by the local authorities. By this or other means he would undertake to secure to the Company, for a period of six months, 6,000 men per month—14,000 less than the present contingent. This last undertaking was to be an unofficial arrangement between the Viceroy and the Porte. The first two were officially embodied in the Grand Vizier's letter,

It contained also a stipulation that the Canal should be used only for commercial and not for military purposes. A commission of engineers was to examine its proposed dimensions with a view to excluding warships—an idle provision, as Palmerston drily minuted, 'for merchant ships are now as big and draw as much water as frigates'. Finally, the Viceroy was authorized to seek an understanding with the Canal Company on these various points and to report back the result to the Porte.

*　　　*　　　*

Ismail had reason to be pleased with Nubar's performance in Constantinople, which he saw as a diplomatic success. The modifications in this second Vizierial letter restored to him much of the political initiative of which the first had deprived him. It gave him room to manœuvre on his own ground in the revision of the Canal Concession. Above all it gave him, for the first time, official freedom to negotiate directly in Paris, implying once and for all the Porte's recognition in principle of the piercing of the Isthmus of Suez. Nubar, on his return, was rewarded for his services with the gift of an estate which had belonged to one of Ismail's wives, which in the previous year had brought in a revenue of 75,000 francs from cotton, and which had since been fully equipped with the latest agricultural machinery.

After only a few days in Egypt, he left for Paris, fully aware of the fact that Lesseps was on his way from Trieste to Egypt and that their respective ships would cross *en route*. In the circumstances Tastu, the new French Consul-General—who, unlike some of his predecessors, was to prove a staunch ally to Lesseps and the Canal—expressed his surprise at Nubar's departure. Nubar replied that he intended to negotiate, not with Lesseps, but with the French Foreign Minister, now once again Drouyn de Lhuys. He brushed aside Tastu's objection that the French Government had not been asked to negotiate, and had refused to intervene with the Company on the points in dispute. Moustier, he declared, distorting the Ambassador's words, had himself advised him to go to Paris.

He left, armed with a letter from Ismail to Drouyn de Lhuys, announcing that he had come to negotiate with the Company, with the Porte's authorization and under the auspices of the French Government, whose good offices and benevolent advice would surely bring 'a prompt and easy solution'. It suited Nubar at the moment to avoid and short-circuit Lesseps, in view of schemes of his own which he had in mind for the further discomfiture of the Canal Company.

Lesseps, on his arrival a day or so later, found that Ismail had left for Upper Egypt. In a Viceregal steamer placed at his disposal, he followed him up the Nile, through the full heat of August in flood-time. He carried with him a long memorandum to the Viceroy, contesting the Porte's new instruction on a variety of cogent grounds. He stressed its illegality in terms of the Company's

Concession, its evident intention to prevent the Canal's completion, and the fact that it compromised the rights of the Viceroy,
transferring the internal government of Egypt from Cairo to
Constantinople, and making a dead letter of the treaties which
governed its status. Meanwhile the Company, adhering to its
agreements, would 'carry on its work with calm and moderation,
but also with the unshakeable perseverance which the Egyptian
Government expects from it, and which every day renders more
powerless the opposition of English diplomacy'.

Lesseps was received courteously by Ismail, but drew from
him, at this and a subsequent meeting, only the statement that
he had confided his intentions to Nubar. Thus no discussion
took place between them. Back in Alexandria Lesseps reported
on his interview in the language, as Colquhoun put it, 'of one
who has gained his cause'. In fact, for all his outward buoyancy,
he was deeply disturbed by the course of events, and had warned
his associates in Paris, by telegram, not to discuss the affairs of
the Company with Nubar. Meanwhile Drouyn de Lhuys, though
he received Nubar with courtesy, made it clear to him that he
could not discuss the affairs of the Canal, which arose from a
freely agreed contract, except at the Company's request. Nubar
must thus await Lesseps's return.

But what most concerned Lesseps was a new and perilous
threat to the Company's existence, which had developed in his
absence from Paris. Soon after his arrival in Egypt he received
a letter from a trusted Parisian friend. This warned him that a
certain French statesman, whose name the writer could not
divulge, had sent a leading engineer in the Government service to
Egypt, to inspect and report on the works in the Isthmus. His
conclusions were highly unfavourable, and his report on them was
likely, through the statesman in question, to be referred to the
Emperor. It was hoped to persuade the Emperor that the direction
of the Canal works was bad, that the shareholders' capital was
being jeopardized, and that the honour and success of the enterprise was in danger. The Empress too was to be invoked, to save
Lesseps, in his own interests, from the embarrassment which he
was thus laying up for himself, and so induce his withdrawal.

The purpose of these moves was to liquidate the Company and replace it by another which was already in the process of formation, and there was talk also of a Company in which the big bankers were interested. The 'statesman' behind this plot was the ambitious and unscrupulous Duc de Morny.

Auguste Morny was the natural son of Queen Hortense, by the Comte de Flahaut, himself a natural son of Talleyrand. He was thus half-brother to the Emperor whom, with his imperial beard, he closely resembled. He had played a prominent part in his *coup d'état* and had since remained a leader of the 'swell mob' in his confidence. For some time Morny served Napoleon as President of the Imperial Parliament, manipulating urbanely the insignificant forces which gathered around the throne. An airy and elegant man-of-the-world, he was portrayed by Daudet, who had been one of his clerks in the Palais Bourbon, as a cross between Richelieu and Beau Brummel.

More exactly he was seen as a bandit in the skin of a *vaudevilliste*. His main role, played on and in the wings of the seething stage of the Second Empire Bourse, was that of a financial buccaneer in the grand manner. 'Morny is involved in the affair' was a frequent assumption in the investment market, generally causing the price of shares to rise. Exploiting his official prestige and his inside political knowledge for his personal profit, promoting from behind rather than from above (and with willing investors to take most of the risk) a series of large-scale speculations, he specialized in international ventures, and it was thus inevitable that sooner or later he should involve himself in the affairs of the Suez Canal.

The Duc de Morny and Nubar Pasha were two of a kind. The Armenian was dazzled by the Frenchman's panache and operational skill, and henceforward they conspired in close concert. It was Morny's intention to bring down Lesseps, and to take over the Canal Company through a group of financiers of his own persuasion, who were already trying to buy up Canal shares— without much success, since they remained resaonably firm and infrequently came on the market. Nubar abetted Morny's intrigues. First Ismail must be deterred from negotiating with Lesseps. Thus Nubar advised his master: 'Continue to act towards

M. de Lesseps as you have been doing. Plenty of politeness, no negotiations. That is M. de Morny's advice. The Viceroy, in deciding to negotiate in Paris, is exercising his rights as a Sovereign. M. de Lesseps must accept this.' Thenceforward, in his despatches, he zealously encouraged the Viceroy in the belief that, as Ismail began to boast to his entourage, he was 'absolute master of the situation in Paris'.

Most revealing was a phrase in the first despatch: 'Persevere in the same line. Principal shareholders favourable.' It was now clear that Ismail shared to the full Nubar's designs on the Suez Canal Company, and his plans for an alternative instrument to finance its completion. His initial swift decision, on returning from Constantinople, to take up his Canal shares and reach agreement with the Company was inspired by no love of Lesseps or loyalty to the Company itself. It reflected in the first place a desire to conciliate France, as the power most likely to further his interests in relation to the Viceregal succession; but in particular a determination to gain control of the Canal, in whose future—thanks largely to Hawkshaw's report—he now fully believed, for himself. This policy had a double advantage. For it might equally serve to placate Britain, who sought also to liquidate the Company and had talked of arrangements to buy it out on Egypt's behalf.

For some months past Ismail had been in confidential relations with Edouard Dervieu, a French banker in Alexandria who generally advised him on his affairs, and who, together with a fellow-banker, Henry Oppenheim, was already, by agreement with the Company, handling at a discount the treasury bonds remitted in instalments by Ismail on account of his shares. Through Dervieu Ismail was keeping an eye on the market in Paris for any fluctuation in the price of Canal shares, in relation to the progress or otherwise of Nubar's current negotiations. He was playing with the idea of acquiring more shares, not merely to gratify his innate speculative instincts, but in the hope of acquiring majority control of the Company. Meanwhile Nubar was lobbying shareholders and others who showed signs of discontent with Lesseps's methods and his handling of the Company's affairs.

Bulwer got wind of these manœuvres, and wrote confidentially to Lord Russell that the Viceroy 'intends in the meantime to become master if he can over the Company and Lesseps by purchasing a large majority of the shares. It appears to me that in this way we shall settle this matter as satisfactorily as we could expect.'[1] Otherwise Ismail was concerned to keep in with any group of financiers or bankers in Paris who might succeed in taking over the Company and afford him, as Viceroy, a greater measure of control over its operations in Egypt.

The success of such a scheme, however, depended on the official policy of the French Government itself. This was still undisclosed. But Ismail was optimistic, taking at their face value hints in Nubar's letters that he enjoyed the support of the Emperor, and of the Foreign Minister, Drouyn de Lhuys, whose courteous rejection he chose to interpret as a sign of encouragement. When the French Consul-General assured him to the contrary, the Viceroy merely smiled 'with an air of incredulous disdain'. He knew better. Morny was involved in the affair, and he had letters also from Morny which gave him every encouragement.

Encouraging too was the attitude of the Parisian Press. Nubar, using Lesseps's own propaganda weapons against him, with funds lavishly provided by Ismail, had launched a press campaign against the Canal Company, stressing principally the evils of its use of forced labour, and this appeared to be making an impact, which extended to London. He was working on the principle that French opinion in support of the Canal had been created by the Press, hence the Press could now reverse it.

*   *   *

Lesseps himself was still in Egypt. He resisted all efforts to persuade him to return to Paris until, still in pursuit of his *fait accompli*, he could make sure of the effective continuation of the works in the Isthmus, where petty official obstacles had once more arisen. He was now concentrating mainly on the study of eventual means to make good deficiencies in manual labour by

[1] Bulwer Papers, July 9, 1860.

increased mechanization. But by the end of September he was able to leave, and soon found himself immersed in the cauldron of intrigue and recrimination which had come to the boil in his absence.

Soon after his return, Nubar officially placed before the Company the Viceroy's proposals for the revision of the Canal Concession. They followed the lines of the Grand Vizier's letter, specifying that the labourer's minimum wage should be fixed at a sum roughly double the previous rates, and that the labour force should be reduced to six thousand.

While the members of the Administrative Council deliberated, Nubar sought to hurry their decision by a personal letter to Lesseps: 'A fortnight', he wrote, 'is little or nothing to us who are living quietly in Paris, but it is a lot to our fellahs. . . . This is not a matter of diplomacy, it is a matter of humanity.' Lesseps rebuked him sharply for these histrionics: 'I understand very well that you should stress this question of humanity with lawyers you consult or with people who do not know Egypt . . . a question which has always concerned me in my life more than it has concerned you in yours.' Such verbal expedients, characteristic of Nubar's technique, were all very well, Lesseps implied, for the public, but to try them on himself in person was going too far.

After due consideration the Council, on Lesseps's advice, categorically rejected the proposals, agreeing only to an impartial enquiry into the average rates of pay prevalent in Egypt, and to a rise in its own rates if they proved lower than those elsewhere. Otherwise, it insisted, the demands amounted to a unilateral denunciation of clauses legally agreed with the Company, threatening to deprive it of essential resources, and by consequent long delays in the completion of the Canal to involve its shareholders in losses amounting to hundreds of millions of francs.

Nubar's reply was to submit a long note to the Foreign Minister, which fell, as before, on deaf official ears. The Viceroy showed his annoyance, but still cherished the belief that he could win his cause through Morny. He now asked the Duke to approach the Emperor on his behalf, and to seek his personal opinion as to the best means by which he could avoid a conflict with the Company. Shortly afterwards Morny invited Lesseps to call on him, and proposed that they should try to reach some agreement between them on the subject of Nubar's mission.

Lesseps at once took a frank line. '*Monsieur le Duc*,' he said, 'You are the last person to mediate in this matter.' He referred to rumours rife in Cairo that the enemies of the Canal were using Morny's name as cover for the Nubar mission, to which the French Consul-General was opposed, and were counting on Morny to ensure its success. 'You must see', he continued, 'that in the face of such rumours caution on your part is called for.' He added that, as intermediary between the Egyptian Government and the Emperor, he could recognize no one but the Ministry of Foreign Affairs.

Morny then revealed that he was not proposing his own intervention. He was proposing the intervention of the Emperor himself. Following the Viceroy's enquiry, the Emperor had asked him to obtain from Lesseps an account of the dispute, backed by the relevant documents, and to submit a report for his consideration. Morny read him a note from the Emperor—which, as Lesseps remembered it, ended: 'You know how interested I am in M. de Lesseps's enterprise, and how much I wish it success. If the Company has complaints to make about damage to its interests, I shall have them seriously investigated, and if it is within its rights I shall see that justice is done to it.'

Lesseps did not conceal his satisfaction. The Company, he replied, would be happy to see the Emperor thus concerning himself with its affairs. Morny replied that he personally was too busy to go through the Company's voluminous dossier and prepare the report which the Emperor required. He proposed instead to delegate the task to a Parliamentary deputy, well known to himself and acceptable to Lesseps. To this Lesseps agreed, and Morny's choice fell on a Liberal leader of the Opposition, Emile Ollivier. The two men parted cordially enough, and Lesseps emerged from the meeting reassured by the turn of events. Once again, in this moment of crisis as in that of 1859, the Emperor was to take a hand in the solution of the problems of the Suez

Canal—this time at the direct request of the Viceroy himself. Lesseps was confident of a successful outcome. And so was Ismail.

*   *   *

Now that imperial arbitration was in principle sure, the Canal campaign, through the remaining winter months of 1863, grew more intense as each side redoubled pressure, skirmishing and lobbying by all means and on all sides for an award which would further its cause.

Nubar, regardless of scruple or cost, waged his press war with increased ferocity, invading even the columns of an organ deemed semi-official. Lesseps, in a speech at Lyon, likened Nubar's campaign against him to that of Palmerston, who 'cried to the English capitalists: "It's an intrigue and a dishonest deception, have nothing to do with the Company!" Now they are crying to the French capitalists: "It's a very bad business . . . clear out of the Company!" They stop at nothing to achieve this. Each one of our successes excites more fury. They manipulate the markets, they multiply articles and deliver copies of them to shareholders at home, they print and hawk around all kinds of false rumours and false news, they spread alarm, they threaten the loss of all the capital engaged. In short, the business must at any price be torpedoed.'

Nubar wrote a report to Morny, for the Emperor's eye, in which he recited at length the Egyptian Government's grievances. He dramatized the 'exploitation of a whole people by a foreign company'; he pretended by means of dubious statistics that the *corvée* in the Canal Zone was costing Egypt millions of francs each year through loss of labour and cultivation elsewhere; he cited a single hypothetical case to prove that the fellah, far from earning extra wages, was out of pocket by the time he returned home at the end of a stint in the Isthmus, where the workers were in any case brutally ill-treated by the Company's employees. Indulging freely in sentiment, he shed crocodile tears: 'I have seen with my own eyes', he declared, 'the sufferings of these wretched and unfortunate fellahs. . . . I have wept, yes I have wept bitterly over these poor compariots. . . . He who strikes a fellah and breaks his wrist with a blow of the *kurbash*, strikes me and breaks my own wrist.' On this Lesseps minuted drily: 'Nubar has never been in the Isthmus. . . . Nubar was Director of the Railway when thousands of Egyptian corpses covered the line . . . on account of the absence of water.'

Turning to the subject of the Company's lands, Nubar minimized their extent and their value to Egypt. He declared that they were nothing but sand, and 'the sand, like the sea, takes a long time to conquer'. Reclamation of them would take thirty years, with the aid of 'financial miracles and an incalculable sacrifice of men'—and even then they would only be fit for the cultivation of barley. Such assertions read strangely to Tastu, to whom Nubar had declared earlier: 'If Monsieur de Lesseps would consent to restore to us these lands, for which we would pay what he asked, we should put at his disposal not 20,000 but 100,000 fellaheen.'

Nubar's report put forward no constructive proposals, and ended with a flavour of blackmail. He recalled a time when the Viceroy Abbas, failing altogether to gain support from the French Government on a matter of life and death for himself and for Egypt, had sent him successfully to Constantinople to obtain it from the British Ambassador. 'For my part', Nubar wrote, 'I sincerely hope that this time I shall not be reduced to return once more to Constantinople and knock at the door of Sir Bulwer.' On this Lesseps minute: 'What impudence!'

The dispute meanwhile reached the law courts. Nubar accused the Company of acting illegally. The Company, on the grounds that he had falsified a document to prove this, accused him of defamatory statements, and brought a case against him in the Tribunal of the Seine. A long legal wrangle resulted. In the course of it, while the case was still *sub judice,* the Company publicized its cause through the Press, following Nubar's consistent example. On this point the Court ruled against it, and neither side was the winner.

In the stock markets there was intensified warfare. Ismail,

abetted by Nubar, was still plotting to obtain control of the Company through an increase in his own shareholding. He hoped that if the Company could be so discredited as to create the general impression that an official arbitration would go against it, the shares could be forced down to a price at which he and certain associates in Paris could profitably enter the market. Then, through the Viceroy's nominees, at an Extraordinary General Meeting of shareholders, convened for the spring, Lesseps could be forced to resign.

Tastu reported the Viceroy's intentions: 'Once this powerful personality has been got rid of, he hopes that the rest of the shareholders will leave the control of the Company's affairs in the hands of the largest shareholder, i.e. the Viceroy.' Ismail's bankers were instructed to buy all they could for him below the market price. But the shares, which were mostly in the hands not of speculators but of small investors, and which seldom came on the market, remained firm, and grew firmer as the tide seemed to turn in the Company's favour.

At the end of the year, Emile Ollivier's report on the dispute, made on Morny's behalf, was ready for submission to the Emperor. Its conclusions did not further the Company's cause. The acquisition of the lands, it argued, represented a separate speculation from that of the Canal and were unnecessary to it. Their retrocession by the Company should therefore be settled in terms of an indemnity, which the Viceroy accepted the obligation to pay. On the other hand the free ownership or control of the Sweet Water Canal was necessary to the operation of the maritime Canal, hence should be retained by the Company.

The possession of the lands depended on a signed legal contract with the Viceroy. The supply of labour, on the other hand, did not. This depended on a sovereign act by the Viceroy, who was free to furnish, at will, either 6,000 workers or no workers at all. M. de Lesseps maintained that the system, as practised in the Isthmus, was not forced labour. But wherever an obligation to work was involved, labour was forced. It amounted to a form of civil conscription, which neither the Porte nor Egypt could be obliged to practice. No indemnity was therefore due to the

Company for the withdrawal of labour. On the other hand, there should be no obstacle to the recruitment of free labour, and the Company should no longer be limited to a labour force in which four workers out of five must be Egyptians.

The Emperor's own views on the dispute were still unpredictable. Bulwer, passing through Paris early in January 1864, had the opportunity of assessing and, so he hoped, influencing them. He had a 'long and interesting talk' with the Emperor, whom he had not met since he was President and whom he now found 'more assured and flowing in conversation, expressing his ideas clearly and confidently yet guardedly'. The Emperor himself raised the question of the Canal. 'Mr. Lesseps', he said, 'is making a great noise about it. . . . I hear you have been over the ground; give me frankly your opinions, and those of your Government as to the whole of this affair.'

Bulwer did so at length, confining himself to the question of the lands and omitting that of the labour. To the Canal itself, he said, there was no objection. 'But whilst the Canal was one thing, the questions, that the Company wanted to connect with it, were another.' In the extent of the Company's present concessions of land, not sanctioned by the Porte, he suggested that 'people would see, not unnaturally even if erroneously, a scheme of French ambition and not of general utility; and that if making the Canal as a universal benefit was the main object, anything that tended to invest it with the character of being made for one nation's particular advantage would, I thought, rather impede than advance its progress'. The British Government, with its interests in India, could not accept with indifference such a state of affairs, nor surely could the Porte or the ruler of Egypt itself.

The Emperor suggested that the Egyptian Government should buy the lands from the Company. 'Exactly, Sir', replied Bulwer. 'That is just what the Egyptian Government has been fairly endeavouring to do, and what Mr. Lesseps is resisting.' He explained that the Viceroy, with the assent of the Porte, had offered him fair and liberal terms, 'but that notwithstanding this, and that various distinguished French lawyers had declared that Mr. Lesseps's claims were illegal, he still persisted in them, and seemed disposed to try to drag his Government and his country into difficulties for the sake of his own pecuniary interests.'

Bulwer then saw Morny, whom he quoted to Russell as saying that the Emperor had assured him: 'I am happy to find that Bulwer and you are agreed and that we all three entertain the same opinion.' The Ambassador's mood was optimistic: 'M. Ollivier's report is against Lesseps; Morny's opinion will be against him, and the Emperor is prepared to agree with Morny.'

The submission to the Emperor of the Ollivier report had been followed on January 6, 1864, by that of a petition from the Company. It begged him to call the attention of his Ministry of Foreign Affairs to the present dispute, with a view to intervention with the Porte for the Sultan's firman on the basis of the Canal's future neutrality. The moment had surely come, it submitted, to resolve this question, 'so that French interests, united together in good faith in an enterprise of general and national utility, should not be compromised by political conflicts'. Pending a decision, the Emperor was further requested, in view of orders in preparation in Constantinople at the instance of British diplomacy, to ensure that there should be no suspension of the works in the Isthmus.

At the same time Lesseps applied for an audience with the Emperor, in a confidential and personal note to his 'guardian angel' the Empress. The purpose of the Company's petition, he explained to her, for the Emperor's ear, was to detach the Duc de Morny from his intervention in the Suez affair, and to hand it over to the Ministry of Foreign Affairs. In Egypt, he wrote, in the present disturbed situation, the British agent's influence was paramount. The French agent's influence, as the mouthpiece of his Government, was non-existent. What counted there was the intervention of important persons, believed to be in the Emperor's confidence. It was, he urged, essential that at this moment in Egypt the unimpaired influence of the Emperor alone should prevail.

The audience which followed was unofficial. But it helped to produce the immediate result for which Lesseps had hoped.

After it, the Emperor directly instructed his Minister of Foreign Affairs, Drouyn de Lhuys, to study the Suez dispute and to recommend Government action. Here was the instruction for which the Minister had been waiting and hoping. It was to lead at last, in the spring of 1864, to an official imperial arbitration on the Suez Canal.

CHAPTER XIII

The Imperial Arbitration

The Quai d'Orsay now worked, through the first months of 1864, to obtain agreement between the contending parties to terms of reference for an Imperial Arbitration Commission. Meanwhile Lesseps's prestige received a well-timed boost from a banquet given in his honour in the great hall of the Palace of Industry in the Champs Elysées, under the presidency of the Company's patron, Prince Napoleon Bonaparte. Familiar to all as the Emperor's fat cousin Plon-Plon, the Prince was the son of Bonaparte's brother Jérôme. A French wit once said of him, unkindly, that he was 'a good copy of the first Emperor dipped in German grease'. In fact, he was a man of varied talents, with much of his uncle's vitality and insight, if so little of his military spirit as to earn from his soldiery, on invaliding himself home from the Crimean War, the additional nickname of 'Craint-Plomb'.

Prince Napoleon was a convinced democrat, who had stood by his cousin from the early days of the Republic, and the Emperor had since loyally helped him out of a number of embarrassing scrapes in which Plon-Plon's public political indiscretions were apt to involve him—scrapes which would usually be followed by spells of observant and intelligent globe-trotting, discreetly away for a while from the disturbed arena of Second Empire France. A born free-lance and a man of his age, with a strong dash of the showman in his nature, the Prince had excelled as director of the Paris Exhibition in 1855. Similarly, showing a flash of Bonaparte's quick sense of essentials, he had come forward from the start as an outspoken protagonist of the Canal scheme, and a warm ally of Lesseps, whose liberal outlook and imaginative vision he shared.

Between them they had chosen an appropriate moment for a public tribute to Lesseps. The Emperor's arbitration was

193

pending; so also was an Extraordinary General Meeting of the shareholders. And at this moment the Company had something new to show them. At the turn of the year, the Sweet Water Canal, linking the Nile with Ismailia, reached Suez. Lesseps, in his speech at the banquet, referred to the 'festival of peace and of labour' which had marked this historic occasion, and described in his own rhetorical terms how 'a river sprang forth across wastes hitherto condemned to an eternal desolation. The Moslem populace ran to the stream of sweet water, plunged their hands into it and moistened their lips with it to convince themselves that it was the blessed Nile. In the midst of his co-religionists a old man cried: "The Christians are also the children of God! They are our brothers!" The brotherhood of all races and creeds was revealed to the astonished crowd, and the same day the wings of electricity spread the news across Europe: "The two seas are joined! The Nile is at Suez!" ' For Egypt, he added, this meant the addition of a province to her rich territory. Her Eastern port had been freed from the grip of the desert to enjoy life and prosperity. History would record this event with the words: 'The nineteenth century willed and accomplished what ancient times had not dared to attempt.'

Meanwhile the corpulent Prince, who had himself toured the Isthmus on one of his tactful disappearances from Paris, treated the guests to a long display of boisterous and down-to-earth oratory. One by one he disposed of the various obstacles now raised by the Company's enemies. They talked, for example, of the rights of the Porte. But what in practice did these amount to? In the Orient, rights were dominated by facts. So Lesseps, with his profound knowledge of its countries, had realized, and so he had acted. The Company held its rights, in fact, from the Viceroy of Egypt, and the Viceroy had in fact assisted its operations for the past eight years—years which would otherwise have produced not one canal finished and another far advanced but merely mounds of paper. How could the Viceroy now turn round and say: 'What I did was wrong. You have spent forty millions of French money. Very well, it is money wrongly spent. I shall stop you and shall arrange with my sovereign to see that you lose it.'

In theory, the Viceroy of Egypt was denied many rights by his sovereign—the right, for example, to condemn a man to death or to make him a Pasha. In practice what he did was to have malefactors unobtrusively drowned in the Nile and to make Generals Beys with the rights of Pashas. Thus rights matched facts and both sides were satisfied.

All that the Prince demanded was treatment for the Canal on a similar basis. In fact, the Company could insist rightfully and without risk on the implementation of its agreement with the Viceroy of Egypt. Nor was there any prospect of serious risk from the opposition of Britain. Enlarging with good-humoured irony on the 'mirage of freedom' which so greatly attracted him across the English Channel, Prince Napoleon warned his audience: 'But quite distinct from the English nation and English opinion there is the English Government.' He produced his own explanation for British policy: 'It is not in the Blue Book, it is in the Peerage. . . . When one opens the Peerage, and one sees that the noble lords now in power are seventy, seventy-five, eighty years old, one understands better why, together with the experience of age, they have hearts which are a little cold. One understands their indifference of feeling towards the most generous causes.' But this did not mean that the British Government would go so far as to drag the British people into a war against a perfectly just cause. 'A war on account of the Suez Canal?' he exclaimed. 'Come, come . . .' The idea was absurd.

All the same, if the Company were ruined by Nubar's present campaign against it, the Canal would still eventually be made. The Egyptians, a people who would lose their trousers rather than sew on a button, were incapable of making it. It would be made, so he predicted, by the capital and labour of Britain, profiting by all that the French Company had spent and done. 'Should we put up with that? Not at any price.' (Prolonged cries of 'Bravo!') Nubar had come to Paris with letters of introduction which were letters of credit from English banks, and with his pockets full of sovereigns not francs. He had failed to undermine French opinion by stealth, and was now trying to influence it by an appeal to noble sentiments. To breach the

o

Company's defences, he had turned to the weapon of the *corvée*.

Here the Prince spoke as a man of liberal principles, who deplored the use of forced labour more than anyone—'more', he added, 'than the Egyptian Government itself'. It was an ancient Egyptian institution which the Company had found and made legal use of and improved upon by paying its workers and treating them well. It was none the less a bad institution. But did the Egyptian Government plan to abolish it? No. It planned to abolish it only for the Suez Canal. On the big cotton and sugar estates of the Viceroy and the Pashas forced labour would continue.

If the Government, in the interests of humanity, was prepared to emancipate its own fellaheen, all credit to it. Let it do so and make good to the Company, in terms of its contract, the additional cost of free labour. This it had already done in the similar case of another French Company, the *Messageries Impériales,* which was at work on the harbour of Suez. But it was not the duty of the Company to play the philanthropist, to bear the brunt of the Egyptian Government's humanitarianism and emancipate the fellah on its own. If the members of its Council agreed to do such a thing, without compensation, they deserved to be handed over to the police or sent off to a reformatory.

With regard to the lands, which the Egyptian Government now sought to expropriate, let the Company negotiate a fair deal with it, in terms not of their actual but of their development value. Why not work for an agreement by which the lands should be ceded to the Viceroy, not now but successively over a period of two or three years, at the prices prevailing when the work of reclamation was finished? To cede them at once, before their value could be fairly calculated, would mean either selling them at a loss or demanding too high a price for them. And this would be folly.

Over these various matters the Company, while still remaining firm, should pursue, in its own interests and those of Egypt and the rest of the world, a conciliatory policy, designed to achieve a just and at the same time a practical solution. If the *corvée* could be abolished and the lands relinquished, both on equitable terms, most of the difficulties inherent in piercing the Isthmus would vanish. Above all, let everything be discussed, frankly and openly. Light was on the Company's side; only darkness was against it. 'Act in full sunlight', he urged. 'Who are these arbiters, what are these counsels and interventions of which one hears so many rumours? I know nothing about them. I don't want to know anything about them. Don't give them a thought. Everything that hides in the shadows, everything that doesn't emerge officially into the light of day is a bad thing.'

This parting shot at the Duc de Morny provoked rousing applause from the guests, which was redoubled and prolonged at the Prince's peroration: 'All I have said is my individual and personal opinion. . . . But I am so deeply convinced of the goodness of the cause I have just defended, and of the justice of the ideas I have expressed, that if public opinion adopts them, I like to hope that the Government will approve them too. I have confidence in the Government of the Emperor, the natural protector of the rights of French citizens abroad.'

*   *   *

A fortnight later the Extraordinary General Meeting of the Company was held in Paris. The shareholders, reassured by such stirring propaganda, and now officially informed by their President of the Emperor's promised arbitration, reacted to the proceedings with 'indescribable' enthusiasm and continuous cries of 'Vive l'Empereur! Vive l'Empereur!' They were informed also that their Council had already handed to the Foreign Minister the Company's counter-project for the settlement of the dispute, for submission through official channels to the Viceroy, and ultimate consideration by the Arbitration Commission.

In this the Company was conciliatory to the Viceroy's demands but firm in its own demands for compensation. The project allowed for the retention of the Sweet Water Canal, as being essential to the working of the maritime canal. It agreed to the retrocession of three-quarters of the cultivable lands, the remaining quarter, a double band along the banks of the Canal, to be retained as being essential for the same purpose. In compensation for this the Company claimed an indemnity amounting to 50 million francs, a sum computed on the basis of the estimated value of the lands when the maritime canal should be completed. With regard to the forced labour, the Company undertook to demand no more than 6,000 out of the 20,000 workers to which it was contractually entitled. The consequent withdrawal of 14,000 workers would create a deficit which would have to be made up from foreign labour, and for this it claimed an indemnity of 40 million francs.

This total indemnity of 90 million francs amounted to a little more than the total of the Viceroy's shareholding. Thus the Company proposed that it should be remitted by the return of his shares, with an issue of Treasury bonds to cover the balance. In addition it insisted on the cancellation of the 15 per cent annual share of the Company's net profits, which had been allowed to the Egyptian Government in view of the concession of the lands and other advantages. Finally, the project was strictly conditional on the issue of a firman which embodied the Sultan's sanction.

It was turned down by Ismail with a contempt which indicated that he was still being encouraged by Morny to except an Imperial decision in his favour, involving a moderate indemnity for the lands and none for the withdrawal of the labour. To Tastu, the French Consul-General, he declared confidently that all was now in the Emperor's hands: 'I cannot accept M. de Lesseps's Note. The time for negotiations has passed. I sent somebody to M. de Lesseps but he refused to negotiate with him. Now it is too late.'

Ismail's own project, involving the reduction of the labour force without an indemnity and the return of the lands, against an indemnity, followed the lines of the Ollivier report, for whose implementation he clearly hoped. But in fact the Emperor, in reply to his official request for arbitration, made it clear that this would cover not merely the matter of compensation for the lands, as the Viceroy had requested, but the whole range of matters involved in the dispute, and to this Ismail incautiously agreed. On this wider basis the Arbitration Commission met on March 18, 1864, under the Presidency of Thouvenel, and after a month of wrangling between Nubar and Lesseps, the two parties finally agreed and signed its terms of reference.

At the last minute there came an objection from the Porte. Hitherto, despite pressure from the Viceroy in Constantinople and from Nubar on the Turkish Ambassador in Paris, the Porte had abstained from intervention with the French Government in support of the Egyptian claim. Nubar, left to his own resources but still hoping for such support, had delayed his signature until he could delay it no longer. Now suddenly the Turkish Ambassador, Jamil Pasha, received instructions from the Porte to object to the terms of reference and to insist, as a prior condition, on the return of the Sweet Water Canal to the Egyptian Government. It argued, with Bulwer's encouragement, that the retrocession of the lands was valueless as long as the water for their fertilization remained under the Company's control.[1]

But the Ambassador, though he received his instructions two days before the signature, took no action on them until four hours after it. As Bulwer reported to Russell: 'He, good man, with somewhat of oriental pride and nonchalance, considered he might wait quietly with his telegram in his pocket till Nubar Pasha called on him. Nubar Pasha, hearing nothing that was positive on the subject, and fancying that he was left alone in the dispute, signed the *compromis*, and only called on Jamil Pasha to tell him so.' All that the Porte, thus caught out by its own indecisive and dilatory tactics, could now do was to claim the right to reject the Emperor's ultimate award on the grounds that it had not accepted its terms of reference. Accordingly the Ambassador was now instructed to withhold from the French Government the contents of the neglected telegram. This was to prove a serviceable loophole, not only for the Porte but for the Viceroy himself.

*   *   *

[1] This proposal was said to have been inspired by Ismail himself, through the banker Henry Oppenheim, the partner of Dervieu, whom he sent to Constantinople at this time in connection with a *Compagnie Internationale de Navigation et de Travaux Publiques*. This was a company formed, with predominantly English capital, to run navigation in the Red Sea, railroads, canals, and other works in Egypt, and possibly, if the Imperial Award went against Lesseps, to take over the Suez Canal Company itself.

Two months were to elapse before the Commission arrived at a decision. Awaiting it with all the patience he could muster, Lesseps reflected, in a letter to Ruyssenaers in Alexandria, on the paradoxical nature of the whole situation: 'Of all the strange things . . . the strangest is this: a Government (that of Egypt) doing everything it can to place itself under the administrative yoke of another Government (the Porte) and using this voluntary subjection to avoid the execution of a contract which ties it to an enterprise it has constituted. It would have been logical enough for the Government (Egyptian) to seek the support of France to resist, in the name of its threatened interests, opposition to the Suez Canal from Constantinople. But the exact reverse has now happened, and it is that which is strange.'

In the Canal zone itself the Viceroy's supply of forced labour, on the former basis, had been extended by agreement until the beginning of June. As the date approached Lesseps began to show signs of impatience at the prolonged nature of the Commission's proceedings. Already the workers were arriving only at irregular intervals; by May their numbers were down to 6,000, and now Ismail was threatening to withhold them altogether.

Behind the scenes Morny and his friends were still as active as ever. Nubar, a master of delaying tactics, was doing his best, through the members of the Commission, to hold matters up. Above all the Company was beginning to run short of funds—and the Annual General Meeting was due in August. Fearing for its credit, Lesseps wrote once again to the Empress, begging for a rapid decision. He wrote in similar terms to Drouyn de Lhuys, objecting especially to a proposal that the Commission should visit Egypt before completing its enquiries, thus causing a further delay. But finally, on June 9, the Commission's report was submitted to the Ministry of Foreign Affairs.

On the issue of principle, the Commission accepted without question the legal validity of the Viceroy's Concessions to the Company. This held him to the recruitment and supply to the Isthmus of labour, which the Company was then obliged to feed and to pay. This undertaking did not conflict with the prohibition under Ottoman Law of the *corvée*, which applied to labour thus recruited but also unpaid. The agreement with the Company was thus a contractual act, which Mohammed Said had implemented, since 1861, with the promise of 20,000 men and the actual average supply of 17,500—a supply reduced by Ismail to some 6,000. As compensation for the loss of this labour, the Company was entitled to 38 million francs, only 2 million less than it had claimed.

The Sweet Water Canal was to be retained by the Company, but only until the completion of the maritime canal. Then it would be restored to the Egyptian Government against compensation of 16 million francs for the cost of digging it and the loss of navigational charges. The Company would cede to the Egyptian Government lands by both canals amounting to 150,000 acres, against compensation of 30 million francs—10 million less than it had claimed—and would retain lands amounting to 50,000 acres, two-thirds of the amount it had claimed. These reductions were largely offset by the compensation on the Sweet Water Canal. Thus altogether the indemnity awarded to the Company, payable in specified instalments over a number of years, amounted to 84 million francs, only a little less than the amount of its original claim and of the Egyptian Government's total investment. An Award in these terms was signed by the Emperor and officially announced on August 2, 1864.

The Imperial Award was an outstanding victory for Ferdinand de Lesseps. After a decade of toil and delay and frustration and struggle against heavy odds, the end of his troubles and those of the Canal was at last in sight—a consummation due largely to his own personal qualities of resolution and patience, his diplomatic dexterity and practical resource, his talent for handling and inspiring associates and workers alike, his dedicated, single-minded commitment to a work of creation. Here, at long last, was an official act of policy, positively committing the French Government not only to the support of the Canal but to the legal vindication, and—in the traditional terms of French political backing for French commercial interests abroad—to the financial salvation of the Suez Canal Company.

Here was a resounding defeat by France of British policy, which yet gave Britain the liberal terms she had sought—the abolition of the *corvée* and the retrocession of the lands. By thus seeking, in its congenital suspicion of French imperialist designs, to dislodge and in effect ruin the French Company, the British Government had played into the hands of the French and, if ruin now threatened anyone, it was not its enemy, the Company, but in the long run its potential ally, the Viceroy of Egypt.

In a spirit of ill-informed confidence and ill-judged speculation, the Viceroy had asked for the Emperor's arbitration. The resulting Award now saddled him with an indemnity as large as his shareholding, but likely to bring him in only a remote, problematical, partial and indirect return on his money. Above all the Emperor had placed him in a humiliating position. As for the Porte, it now found itself placed in the position of having, at long last, to move towards positive action, in terms of a firman to authorize the Suez Canal. Here, at Constantinople, in its hesitant hands, was one last card for the British and Egyptian Governments to play, and they played it, for what it was worth, without hesitation.

Lesseps, reporting to his shareholders on the Imperial Award, declared: 'We can now give ourselves up to the work without distraction. Its completion is now assured, with little delay, and with the certainty of substantial advantages to shareholders.' He spoke with calculated optimism. The victory was certainly in prospect; but it had still to be finally won, against a protracted and obstinate rearguard action inspired by Bulwer, supported by the Viceroy, and fought largely in Constantinople.

*   *   *

Ismail, having asked without reference to the Porte for the Emperor's arbitration, could hardly do otherwise than accept his Award. But to save face he chose to stress that he did so only as Viceroy of Egypt and could not answer for the Porte's agreement. Nubar was incensed at his betrayal by Morny. Nor was he greatly mollified when Morny, on the Emperor's behalf, invested him with the insignia of a Commander of the Legion of Honour. Bowing low to receive the collar from the shorter man, he is reported to have said: 'Eighty-four million francs around my neck is quite a heavy enough weight to make me stoop.' He too saw the Porte as a possible face-saver and proceeded, on Ismail's instructions, to Constantinople.

Here he found, once more, a ready ally in Bulwer. The British Ambassador had from the outset opposed any Imperial arbitration at Ismail's request as an abuse of the Sultan's sovereignty, which amounted in effect to the transference of Constantinople's authority over Egypt to Paris. He now found the Award to be 'artful and unfair'. It was moreover anomalous, he argued to Russell, that one sovereign among the five or six guarantors of the Ottoman Empire should grant by his own dictum a portion of its territory to his own subjects without its consent or that of the other guaranteeing powers. If Britain allowed this she would earn the contempt of the French nation. Meanwhile the Sultan's Ministers were 'embarrassed as to how to get out of the corner into which the French Government has so unceremoniously pushed them'.

In London the Turkish Ambassador confessed to Lord Russell that the Porte 'were at a loss to know how to act and what course to take'. Russell, rubbing in the fact that the Porte had been aware of the Viceroy's request for the Emperor's arbitration, 'did not see how this award could be well disputed so far as the pecuniary indemnity was concerned'. On the other hand the Sultan might well seek modifications regarding the lands to be retained by the Company, enquiring into their extent and assessing their significance in terms of his sovereignty. It had in any case been recommended, in a covering note to the Award, that a Mixed Commission should be appointed to delimit the necessary lands on the spot, and Russell proposed that the Sultan might fairly insist on its appointment. It now became British policy to press for the Commission. On the basis of its findings the Emperor—so Russell suggested to Bulwer—'when he sees the injustice of the present proceeding will probably give his consent to a moderate, equitable and final adjustment'. Meanwhile the Porte proposed the drafting of a Convention to be agreed between the Viceroy and the Company in terms of the Award, as a basis for the granting of the Sultan's firman.

Lesseps accordingly drew up a draft Convention on behalf of the Company, based on the Emperor's Award. Nubar, however, blandly disregarding his setback in Paris and profiting by Bulwer's encouragement, had resumed negotiations in Constantinople as though—so it was suggested to the Quai d'Orsay by Tastu—'the Emperor's judgment had never been given'. He now drafted his own Convention, to which the Porte agreed.

Having provided for the delimitation and reduction of lands by the Commission, it proceeded to go back beyond the terms of the Award to the terms of the original concessions on which it was based. It proposed radical changes, designed to increase the Viceroy's hold over the Company. The Egyptian Government was to nominate its President; to have a say in fixing its Canal dues; to supervise its works of construction on both canals; to exercise jurisdiction in all matters affecting it; to add to its list of Founder Shareholders. Finally it insisted that the Egyptian Government, having previously been entitled to 15 per cent of the Company's profits, was now entitled to retain 15 per cent of the indemnity allowed to it on the Sweet Water Canal. The proposed Convention reduced the Emperor's judicial and equitable Award to what Tastu called a kind of 'august simulacrum', before which men raise their hats politely, then proceed about their business.

Having thus opened for his master the door to the Porte, Nubar returned with the news to Egypt. Lesseps arrived two days before him, with his own proposed Convention, strictly following the lines of the Award, in his pocket. The Viceroy raised few objections to it, but preferred to delay signature until Nubar's arrival. When Nubar arrived, he handed the Porte's Convention as he defined it, to Lesseps, who read it through, shrugged his shoulders and dismissed it immediately as unworthy of serious consideration. None the less Ismail, with Nubar behind him, invited Lesseps to sign it. This he refused to do, making it clear that the Company would accept nothing but the Emperor's Award, to which the Porte's Convention was totally opposed. Any special Convention was really, he argued, superfluous. All that was now required of the Viceroy was that he should write to the Porte, requesting the Sultan's firman, and Lesseps outlined terms in which such a request could be drafted.

The Viceroy refused to do this. With continued encouragement, it seems, from friends in Paris, he clung to the belief that there was still a bargain to be driven, and made a last obstinate stand on the principle: 'The Canal belongs to Egypt, not Egypt to the Canal.' The outcome was, for the moment, a rupture with Lesseps, and Ismail cancelled a visit which he had planned to make to the Isthmus.

Lesseps now wrote to the Emperor himself, respectfully begging him to instruct his Ambassador at the Porte to demand the firman, as the immediate consequence of his Arbitral Award. The matter was thus referred back to Constantinople, where the Ambassador, Moustier, pressed for the firman with spirit and firmness. But the Porte, in its turn, referred him back to the Viceroy. It took the line that without his request, hence without his prior agreement with the Company, no firman could be granted. The delaying action which arose from this attitude was to continue for more than a year. In the course of long and tedious negotiations, three more successive Conventions were to be drafted before a formula was found to which the Sultan could be advised to grant his consent.

*   *   *

Meanwhile the British Government, fearing the results of French pressure in Constantinople and anxious to reach a solution, decided that the moment had come to intervene with the French Government directly in Paris. Hitherto Britain, in the course of her ten-year offensive, had avoided any such confrontation, preferring more indirect tactics in its endeavours first to defeat the Canal altogether, then to defeat its execution by the French Canal Company. Now that both these battles were lost, only one course remained to her, a last-ditch sortie planned to mitigate the worse effects of the second defeat. This concerned the extent of the lands in the Isthmus now granted for the Company's retention.

Palmerston himself intervened, proposing the grant to it of 'such smaller quantity as may on proper enquiry be found to be necessary for the purposes of the Canal but for no other purpose'—thus tacitly admitting that the Canal, which he had declared to be impossible, now existed as a fact on the point of accomplishment. These were among his last words on the subject, for Palmerston died a few months later, in his eighty-first year, spared the full bitter taste of the fruits of his policy's failure. It was a policy which had served but outlived its time. Over the past few years it had become increasingly evident that a Suez Canal would materialize, a new line of world communications in which Britain would have no direct share. This was to call, in the interests of her overseas Empire, for a reappraisal of British policy, with adjustments and reorientations and a new range of commitments in the Mediterranean and Middle Eastern worlds.

Now Russell, still Foreign Minister but destined soon to be Prime Minister in Palmerston's place, expressed the view to Lord Cowley that the Emperor should appoint a Mixed Commission, as already discussed, to inspect the lands on the spot and agree on their proper delimitation. A new emphasis had been given to the question by the fact that the Porte, as originally suggested by Russell, had sent a Commissioner to the Isthmus, Osman Nury Pasha, to make his own estimate of the extent of land required for this purpose. The amount which he recommended for the banks of the maritime Canal was only one-quarter of the total granted in the Arbitral Award, while his estimate for those of the Sweet Water Canal was a mere 1,500 instead of 24,000 acres. British suspicions were thus fully aroused that it was the Company's intention to use the surplus land for colonization and so, as Cowley expressed it, 'to indemnify M. de Lesseps for losses which he had sustained in persisting in a ruinous speculation'. This was an intention which must at any cost be resisted, more particularly as it infringed the Sultan's sovereign rights as 'lord of the soil'.

Thus Cowley embarked on a series of talks with Drouyn de Lhuys. Drouyn sought to dismiss Cowley's apprehensions as 'mere chimeras'. But he incautiously remarked that shares in the land would be available to anyone, not merely to French nationals

—thus admitting, in the eyes of the British Government, that, despite the Emperor's repudiation of any speculative use for the lands, colonization was in fact intended. Drouyn, no personal supporter of Lesseps since the days when, as Foreign Secretary, he had sent him to Rome on the diastrous mission which ended his diplomatic career, talked frankly and indeed indiscreetly to Cowley.

'Let the Emperor's award be carried out', he said. 'You will then see that the enterprise will assume proportions too gigantic for success. The subscribers to the Company will become clamorous for their money. M. de Lesseps will be obliged to wind up his affairs. A compromise will then become possible, and the Viceroy may repurchase the whole affair at a moderate price, since the subscribers will sooner lose half than all.' In other words, as Cowley read his remarks: 'Make the concessions in order that the Pasha may be mulcted for a further sum of money for Lesseps's advantage.' Drouyn made no secret of the fact that he and the Emperor were now 'heartily sick' of the whole business. He was clearly reluctant to be troubled with it, or to trouble the Emperor with it, further, and at their next meeting he produced to Cowley an 'appeal *ad misericordiam*' rather than a serious argument.

Serious arguments were none the less elaborated in an exchange of Notes, forcibly representing the two Governments' views, while the Porte was itself stirred to action. This diplomatic engagement between British and French political interests was fought largely at the higher level of sovereign rights and dynastic prestige, Russell insisting that 'while the honour of the Emperor of the French is so strangely and improperly brought into discussion, the honour, dignity and independence of the Sultan are truly and seriously involved in the compact now to be made'. Honour was finally satisfied in a compact confirmed in direct correspondence between the two Sovereigns.

Granting an audience to Cowley, the Emperor gave the assurance that 'he would have no objection to the immediate appointment of a Mixed Commission on the lines proposed, 'provided the Sultan should engage, after the enquiry should be terminated, to issue the necessary firman for the recognition of His Majesty's award'. The Sultan wrote to the Emperor, accepting in principle his Arbitral Award but proposing the Commission as a means of avoiding all possible error and eventual conflict. He undertook to accept its decision, whatever it might be, and as soon as he received its final report, to issue his firman authorizing the Canal. Thus the matter was settled.

Drouyn, in despatching what he hoped would be his last note on this subject to Cowley, added a personal postscript: 'Bulwer and Lesseps are embracing on the banks of the Canal. Here we can at least give each other *"une* shake hands".'

*   *   *

Bulwer had indeed transferred his headquarters to Cairo, where he remained for some months, ostensibly 'for the sake of his health', maintaining contact with Ismail and keeping a sharp eye on Lesseps, who was fighting for his Company's rights to the end. Lesseps, with his customary showmanship, took him on a personally conducted tour of the Isthmus—'with all the appearance', as Bulwer put it to Russell, 'of travelling through a French Colony'. The Ambassador, who had not visited the site for two years, admitted progress, but still had his reservations.

Ismailia he found less large and flourishing, and its buildings fewer and less permanent than other travellers had led him to expect. Port Said, on the other hand, had grown considerably, and its buildings were of more solid construction than before. The whole fate of the Canal, he judged, still depended on whether or not a good harbour could be made here, and on balance he was inclined to believe that it could. Independently of the Canal, the port could at least be useful as the start of a hundred-mile road or railway to Suez, halving the distance from Alexandria and now plentifully supplied with fresh water, with fertile lands around it. But he still doubted whether the Canal would ever, 'without great changes in the present scheme, too vast to contemplate, be fit for the passage of large steamers'.

Lesseps, at Port Said, did the honours at a dinner given by the Canal engineers to the officers of a visiting British warship, toasting the Ambassador with hopes of a day when not only France but Britain would have port establishments here, in friendly rivalry and mutual understanding. Bulwer, in reply to the toast, hoped in his turn that the great work would prove genuinely universal, free from any one personal or national interest. More specifically, he advised his audience not to irk the Egyptian Government by demanding of it more than was strictly required. Not long afterwards he relinquished, for personal reasons, his post at Constantinople.

Ismail was now once again on good terms with Lesseps, since he had punctually honoured, in the act as in the observance, the Imperial Award, paying to the Company the first instalment of the indemnity due from him. Not long afterwards the Duc de Morny died, earning the epitaph from Cowley: 'He would have been a great man had he been an honest one.' Ismail thus lost his best French ally. His death, moreover, compromised Ismail with the Porte, which came by certain letters he had written to Morny, and found in them hints of disloyalty between vassal and sovereign. But it meant an end to backstairs diplomacy and to conspiracies for speculative gain. It weakened Nubar's position with Ismail, who became more disposed towards conciliation.

In this spirit negotiations proceeded, seeking agreement on the other problems, apart from that of the lands, which still required to be settled, between Viceroy, Company and Porte, as a final basis for the granting of the Sultan's firman. The thorniest of these was the juridical problem. In his fear, shared by the British, of a French occupation of Egypt by means of the Isthmus, the Viceroy, though he admitted the rights of the Company to direct and administer its own internal affairs, held that legally it should be subject to the jurisdiction of the Egyptian courts. This Lesseps firmly resisted, well aware of the shortcomings, both in principle and in practice, of Egyptian justice; but French official intervention produced an acceptable compromise.

This, with the remaining points at issue, was embodied in two successive Conventions, dated respectively January 30 and February 22, 1866, and duly signed by the representative of the Company and the Viceroy. They provided that legal disputes within the Company should be subject to the French courts in Paris; that disputes between individuals, of whatever nationality, should be subject to local tribunals, in terms of the Capitulations which already covered the rights of foreigners throughout the Ottoman Empire. On the other hand, disputes between the Company and the Egyptian Government should be subject to Egyptian law.

The Egyptian police would ensure law and order throughout the Canal zone, where the Government reaffirmed the right to occupy sites for administrative and above all strategic purposes, thus confirming its freedom to erect fortifications for the defence of the country. It should also have the right to attach a special Commissioner to the Canal Company, to ensure the execution of their mutual agreements.

These agreements were recapitulated in terms of the original Concessions, with modifications embodying those of the Imperial Award. Among the clauses reaffirmed was the right of private individuals, with the consent of the Egyptian Government, to establish themselves on the lands, in the towns and along the banks of the Canal, to be retained by the Company—a significant disclaimer of any exclusive rights to its use, which went far to meet points as to its proper extent now under enquiry by the Mixed Commission.

The agreement included a new and important concession by the Company—the re-sale to the Egyptian Government, for 10 million francs, of the Wadi Tumilat, which the Company had bought from it five years earlier for 2 million francs, and had transformed into a fertile area of some 25,000 acres, now bringing in a revenue of 200,000 francs per year. The Viceroy, with British encouragement, had pressed for this Land of Goshen, where Joseph had settled his countrymen, afterwards to be evicted by the Pharaoh; and where Lesseps was now establishing a similar settlement, largely of Beduin—still no lovers of the fellaheen, hence liable, as of old, to give trouble to any Egyptian Government. Ismail needed it besides to round off the domain which he was creating by the rapid completion of the Sweet Water Canal from Cairo to the Wadi—ironically with the aid of a huge supply of forced labour, and under the auspices of Nubar, who was now his Minister of Agriculture. Finally, in return for its various concessions, the Company was granted speedier payment of the Viceroy's indemnity, providing it with immediate funds which it now urgently needed.

At the beginning of March 1866, the Mixed Commission completed its survey, harassed in the process by the new British Consul-General, Colonel Stanton. He had been chosen to replace Colquhoun on account of his professional experience as an officer in the Royal Engineers, and he contested the Commission's operations in the Isthmus, yard by yard. In its final report, none the less, it recommended the deduction of a mere 150 acres from the total area of land allotted by the Arbitral Award. Only its distribution was materially different, and Stanton was able to console himself, in reporting to London, by the fact that relatively far less land was now allotted to such centres as Ismailia and Kantara. This applied equally to the port of Suez, over which a separate and even more heated dispute had arisen between vital British and French shipping interests. This too was now satisfactorily settled.

*   *   *

On March 19, 1866, the Sultan signed a firman authorizing the construction of a Canal between the Red Sea and the Mediterranean—an authorization for which the Viceroy of Egypt had applied more than eleven years earlier. Such was the measure of delay which a great European power, mistrustful of another, had contrived to impose on the realization of a 'great enterprise', as the Sultan's preamble now described it, 'destined to furnish new facilities to trade and navigation' in 'this century of light and of progress'. Britain did her best to maintain that her opposition had in fact achieved its purpose: the abolition of forced labour, the reduction in the concessions of land, the defeat of plans for French colonization, the maintenance of Egypt's jurisdiction and of her right to construct fortifications.

But these were all questions which might have been resolved

earlier, had any British Government been blessed with the scientific vision to see the Canal's significance instead of declaring it impossible, and the political acumen to share in its benefits instead of opposing it. In a more equable political climate, the Canal project could have beome an instrument of peace, a cohesive rather than a divisive force for a true alliance between Britain and France. But the time for such an Entente was not yet ripe. Instead, the Canal was shunned as a possible new instrument of war between age-old enemies, a chimera which still lived as a reality in the minds and memories of such leaders as Palmerston. The result was a setback to the interests of Britain, and a blow to her international prestige.

She accepted her political defeat with a good enough grace. Lord Clarendon, the Canal's earliest official enemy, was now back in office as Foreign Secretary. He wrote to the Viceroy, earnestly hoping that 'the position in which the affair of the Canal is now placed will be firmly maintained by Your Highness', conveying the thanks of his Government for 'the various friendly acts displayed by Your Highness towards Her Majesty's subjects', and assuring him of its earnest desire to maintain the present cordial relations which so happily existed between them.

France had emerged from this decade of Suez strife with her prestige enhanced. Alive from the start to the possibilities of the Canal and its immense geographical significance, the French Government was yet more concerned with the overriding necessity to maintain and strengthen, as a basic principle of its European and Middle Eastern policy, the Anglo-French alliance. Here, for France, lay the reality. But Britain mistrusted France, and the mistrust became mutual, provoking French fears of aggression by the 'man of 1840' to match British fears of aggression by the heir of the Bonapartes. In this atmosphere of shadow warfare the Emperor feared to provoke Britain over the project for a Suez Canal, which represented for him neither imperialist designs nor political ambitions. Thus he shrank, for a decade, from giving it that resolute official support which might, at any moment, have decided the issue. He delayed it, and thus delayed the Canal's construction, until French interests seemed seriously threatened. Then his intervention was conclusive, a French Suez Canal was assured, and France, at the end of an unduly prolonged diplomatic war, which stopped short of the battlefield, could justly claim a resounding victory.

The Sublime Porte emerged from its passive part in the struggle with its prestige unimpaired. Never greatly concerned, one way or the other, with the merits of the Canal as such, it had directed a long spell of masterly inactivity to the objective of keeping its balance between Britain and France, its two contending guarantors and protectors. Showing a judicious bias towards Britain, as traditionally the stronger, moreover the readier to curb any separatist trend on the part of Egypt, the Porte had yet contrived not to antagonize France. A hastier and more positive policy might have driven the Viceroy of Egypt, with French support, to act, in the matter of the Canal, independently of the Sultan's firman. But he had refrained from doing so, and so the point was reached at which the firman became essential to its final construction. Thus the sovereignty of the Porte over Egypt was confirmed to the world as though it had never been called in question.

If there were a loser in this battle, it was the Sultan's vassal, the Viceroy of Egypt. Ismail, unlike Said, wanted the Canal less for its own sake than as a means of extending his power. Under the firman he had established his Viceregal prerogatives, affirming his internal jurisdiction and his rights of defence, increasing, at a price, the extent of his Viceregal lands. But he had failed to achieve his real objective—an increased independence of the Porte. Hoping for this, in terms of Egypt's international prestige, he had chosen to conciliate France rather than Britain. But to achieve it he must, even at French expense, conciliate also the Porte. Hence Ismail's ambivalent policy, now scorning, now deferring to the Porte, now resolute, now evasive in support of the Canal, and throughout his dealings inspiring mistrust in potential friends as in enemies.

Such contradictions of motive underlay the spirit of vacillation and subterfuge inherent in Ismail's tactics. They were aggravated by his excessive reliance on unreliable counsellors like Nubar and Morny, who played on his ambitions, led him astray into dreams of increased power and fortune, and caused him to obtain less than he hoped through demanding too much. In the end he had obtained, at an absurdly disproportionate cost to himself, part of what he wanted from France. But he had obtained nothing of what he wanted from the Sultan. Nor was he to obtain it in full until seven years later—and then only at a further ruinous cost in terms of bribes and gifts.

Meanwhile, on August 1, 1866, Lesseps could at last justly say to his shareholders that the Canal was now 'no longer a hope but an incontestable fact'. All that now remained to be done was to complete its construction.

CHAPTER XIV

The Canal is completed

Through these years of contention and crisis, from the accession of Ismail in 1863 onwards, the works in the Isthmus sometimes languished, but never wholly ceased. The four earlier years had been a time of preparation, at the end of which Hawkshaw pronounced his favourable verdict on the progress of the work and its ultimate prospects. Reporting early in 1863, he anticipated that the Canal could be finished within five years at a cost of £10 million (£2 million more than the previous estimate), and made light of the various obstacles to its operation.

During this preparatory period the work was more or less confined to the northern stretch. Lake Timsah was reached from the Mediterranean through a shallow maritime channel from the North and by the fresh waters of the Nile, through the Sweet Water Canal, from the West; while progress was made with a deep narrow cutting, down to water level, of the high sandy ridge of el Guisr.

During 1863 the excavation of the maritime Canal proceeded slowly southwards from Timsah towards the Bitter Lakes; a start was made on the cutting of the rockier ridge of Chalouf, between Suez and the Bitter Lakes; and at the end of the year a branch of the Sweet Water Canal reached Suez. This was a year of transition between the preliminary works and the larger-scale works which lay ahead, between the period of manual labour and the period of increased mechanization. Throughout it the Viceroy continued to supply labour, and there was no serious interruption in the work.

By the end of the year, the harbour of Port Said was in good working order; the workshops throughout the Isthmus were well equipped; the main stations were growing into towns; there were stretches of a navigable channel of some kind—if only a few feet deep and for flat-bottomed boats—along much of the line of the two Canals; the lands along the banks of both were becoming settled and cultivated; above all, there was fresh water everywhere, disposing once and for all of the spectre of thirst in the desert. But the bulk of the work still remained to be done.

By now the former single contractor, Hardon, had retired from the scene, since the task had expanded beyond his scope. Instead Lesseps invited tenders from firms of contractors in Europe, who visited the site and were duly impressed. He divided up the various tasks among four of them. Dussaud, who had been responsible for the harbour works of Cherbourg, Marseilles and Algiers, took over the construction of those at Port Said; Couvreux, another Frenchman, supervised the excavations at el Guisr; Aiton, a Scotsman, was entrusted with the first stretch of the Canal from Port Said to Lake Timsah; Borel and Lavalley, two notable French engineers, with the stretch from Lake Timsah to the Red Sea. But Aiton, British and slow and sure in his methods, found it impossible to work with the French Company, whose officials were always in a hurry and over-anxious, in his opinion, to catch the public eye at the expense of sound engineering methods. He fell moreover into financial difficulties when forced labour was no longer available, and at the end of 1864 his contract was cancelled, with a fair settlement from Lesseps which was not, however, to save him from bankruptcy. His task was taken over by Borel and Lavalley. The construction of the Canal was now wholly in French hands.

The supply of forced labour had started to dwindle from the earlier part of that year, and was employed for the last time on the cutting of the channel through the ridge of Chalouf. In June, in anticipation of the Imperial Award, it ceased altogether. Thus 1864 was a lean year for the works in the Isthmus. Up-to-date machinery on a large scale had been ordered through Borel and Lavalley. But it would not be available until the following year, and the existing supply of dredgers was limited.

Meanwhile Lesseps judged that it would be useless to count on voluntary Egyptian labour alone. Thus he launched a recruiting campaign for paid workers from all parts of the Mediterranean, to be engaged under contract. Arabs from Palestine, Syria and North Africa were now to form the bulk of his labour force for manual and spade-work. The European contingent was much increased. Italy provided masons, carpenters, smiths and miners— a number of them political refugees; France a larger quantity of mechanics, dredgers, engineers; while the boats and barges were mounted and manned and maintained by Greeks.

Conflicts inevitably arose within this polyglot community, chiefly between the Arab workers and the lower-grade Europeans. The police could not always control them, the Egyptian administration was ineffectual, and Lesseps had often to restore order by the force of his own personal authority. Once, in 1864, as he was on the point of leaving Alexandria by boat for France, he was recalled urgently to Ismailia, where twenty men had been killed or severely wounded in a pitched battle, provoked by a large group of fellaheen against a small group of Greeks. Only he, with his own loyal henchmen, could make and keep the peace, and this he did by promising the Greeks redress, demanding authoritative measures by the Egyptian Government, and causing it to send to the Isthmus a larger force of police and troops.

In the following year, at an intricate moment in the Paris negotiations, a disastrous epidemic of cholera brought Lesseps hurrying back to Egypt, whose inhabitants he found in a panic. In the Isthmus he mobilized his medical services, himself working among the doctors and nurses, and upheld morale among his employees by a calm and courageous demeanour. This was in notable contrast to that of Ismail, who had fled to sea at the start of the outbreak—an example followed by a third of the population of Alexandria.

As time went by, the new labour force settled down well enough. Gangs were formed, as far as possible, of men of the same nationality. Tasks were allotted largely on a basis of piece-work, the foremen measuring in advance the quantity of earth to be removed and naming the price to be paid for it, regardless of time. Speedy remuneration proved a powerful incentive to zeal, and indolence by any one member of a gang was discouraged by the rest, since it delayed the completion of the work at their expense. At first there was some difficulty in persuading the Arab workmen to use wheelbarrows. They preferred their familiar baskets, containing only a few handfuls of earth, which they often carried on their heads; or sacks, which one would fill and another would carry away. They slept largely in the open, wrapped in blankets, sheltered by two planks running to a point and called, by the French, '*bonnets de police*', which they distinctly resembled.

*   *   *

In the spring of 1865 Lesseps's showmanship was seen at its best with a visit to the Isthmus of nearly a hundred representatives of the principal Chambers of Commerce of Europe and America. Landing in Alexandria, they were surprised to find, among its business community, a strong element of opposition to the Canal as a 'chimerical enterprise'. The main reason for this was a fear that Port Said would replace Alexandria, by diverting much of the trade from the city. In Europe the delegates had been regaled with the usual tendentious rumours, including one to the effect that there was still no water at all in the Canal.

They were thus agreeably surprised by all they now saw for themselves, moreover enchanted by the imaginative hospitality of this man 'who pierces a desert with as little concern as though he had traced a garden path'. At a banquet, where the flags of all nations were draped around a transparency of the Isthmus of Suez, the toast of the Canal was proposed by Cyrus Field, from New York. He was about to inaugurate the first submarine cable across the Atlantic Ocean, and he wished all success to its forth-coming union by Lesseps, for the good of humanity, with the Indian Ocean. The Suez Canal, he declared, would be a monu-ment to his talent and energy as enduring as the Pyramids. Well briefed with Biblical allusions, he added that he was much in sympathy with Moses for electing to reside in so delightful a land, and only regretted that his business engagements did not permit him to do so for longer himself.

After visiting Ismailia, by the route of the Wadi Tumilat,

the delegates were saluted on the ridge of el Guisr by the piercing hoots of locomotives, and entertained to an open-air luncheon on the site of the Canal station, where a garden was striving to bloom amid the sand-dunes. Lesseps now christened the adjoining station with the name of Mariam, in honour of the passage into Egypt this way of the Holy Family—a journey consecrated already by a chapel to the Virgin Mary, built in proximity to the Company's own mosque. Mariam's godmother, he announced, was to be another Mary, the charming wife of a delegate from San Francisco, to whom Lesseps had taken a fancy: 'Thus does America come to the boundaries of Africa and Asia to shake hands with the workers of Europe.'

North of the ridge, the Canal was already a long straight waterway, with banks like a shelving sea beach, enlarged to its full width but not yet to its full depth, and stretching for fifteen miles to the horizon and beyond to Kantara. Here a ferry now plied across it from Egypt into Syria—from Africa, in fact, into Asia. Kantara had grown into a substantial town, nourished by the Sweet Water Canal and enlivened by an expanding caravan trade. Thence Lesseps's party proceeded across Lake Menzaleh, where the liquid mud so laboriously scooped from the bed of the channel had been dried by the sun into hard, straight 'macadamized' banks.

Such embankments were now beginning to line other stretches of the Canal, built up on either side of it in the process of excavation and, as dredging proceeded, given a firm surface by the harder clay soil from the lower levels. They appeared to be effective breaks for any moving sand which might threaten to choke the waterway, and would be more effective still when irrigated and planted to hold the sand on certain stretches, as the proposal now was, thus confuting those pessimists who had foretold such a hazard. Even more telling in confutation was the fact that the excavations, for the Sweet Water Canal, of stretches of the original Canal of the Pharaohs, disclosed intact its original banks. Its bed showed little trace of blockage by sand, despite its abandonment eight hundred years before.

Hawkshaw had disposed, in his report, of this threat from

moving sands, considering that it could easily be met by a permanent system of dredging. The further anticipated hazard of wear and tear to the banks as a result of wash was being carefully studied. Its effects were more noticeable on and near the water-line than at a lower depth. Thus the slopes were being inclined at 1 in 5, diminishing to 1 in 2. It was found too that a natural growth of seaweed often helped to consolidate the banks.

When the delegates reached Port Said, the news was received, and announced by Lesseps, of the death of Richard Cobden, 'the apostle of free trade'—but not, in fact, of the Suez Canal—to whose widow an appropriate message of condolence was sent. In the course of its tour the party covered the Canal site, from one end to the other, in twenty-seven hours' travelling time. Over the northern half of the Isthmus, between Lake Timsah and the Mediterranean, they travelled by water, for the last stage in a small steamer but otherwise largely in flat-bottomed barges, towed by mules and camels. Over its southern half, between Lake Timsah and the Red Sea, they travelled by land, largely on horseback.

Timsah, which Lesseps had first seen as a gaping void of sand and gravel, was now gradually filling, to become a broad stretch of water, if not yet sufficiently deep to accommodate 'the Fleets of the World', as he had predicted. But the Bitter Lakes were still dry, and he now rode with a party across their basin, from which conical mounds, white with salt in the sunshine, rose like tombs to give it the air of a cemetery. The discovery in the basin of sea-shells, like those in the Red Sea, seemed to prove that the lakes were once a gulf of unequal depth at its northern end. Lesseps held the view, which he retailed to his companions, that the Israelites might well have crossed it here, where they were riding. 'The Lord', so the Bible recounts, 'caused the sea to go back by a strong east wind, and the waters were divided, and the children of Israel went into the sea upon dry ground.' The water, partially dried up at low tide, could have become impassable once more at high tide, with a southerly change in the wind, 'when the sea returned to his strength as the morning appeared'—overwhelming the hosts of the Pharaoh.

Hence Lesseps conducted his party to Suez, the insignificant

Egyptian village which was soon to be transformed, thanks to the arrival of the Sweet Water Canal, into a large and active port. As a natural harbour it demanded no works on the scale of Port Said; the Company's activities were limited to the construction of a mole, at the future mouth of the Canal, as a break against southerly gales and high tides; the dredging of a channel from the mouth to the roadstead; and the reclamation of land with the use of the dredgers.

Before the delegates dispersed, the optimistic announcement was made to them that the Company hoped to complete the Canal by 1868, within the limits of its existing capital and of the funds supplied by the Viceroy's indemnities. They were also assured that, with the completion of the sluices on the Sweet Water Canal, and thus the release of more water in the direction of Suez, regular navigation for merchandise from sea to sea would soon be an accomplished fact; and this promise was fulfilled a few months later.

On August 14, 1865, amid rapturous applause and the firing of salvoes, and with the blessing of a Mass and *Te Deum* at Ismailia, the first through convoy of barges passed southwards on its way from Port Said to Suez. Dressed with the flags of all nations, they carried a cargo of coal, 'this raw material which the genius of man has transformed into the world's main instrument of civilization'—and which would henceforward be available at Suez for fifty instead of ninety francs per ton. Its passage from one sea to the other marked the Emperor's approaching birthday, and a telegram of greetings was sent to him, announcing the historic event. A personal reply, with appropriate congratulations, was received from him that very same evening. The workers in the Isthmus were duly 'electrified', as their President put it, by all these manifestations of Progress.

*   *   *

A fresh impetus was given to the Canal works in the course of the summer of 1865, when the arrival and assembly of the new machinery led to their resumption on a more intensive scale than before. When Bulwer, with Colquhoun, visited the Isthmus in the spring, they found 'a large number of ingenious machines', some already in use, some awaiting transportation southwards by the service channel to the various sites on the Isthmus. They included enormous dredgers, cranes and 'new machines called excavators, which worked on the dry sand'.

Thanks, indeed, to the resource and enterprise of Borel and Lavalley, engineering plant of an entirely new kind was invented and, from now onwards, brought into use along the whole length of the Canal for the final stage in its construction. Here was the most powerful dredging machinery yet known to man. Steam took over, widening the stretches of channel which manual labour had dug down to water-level, performing and completing its other tasks with incomparably greater speed and efficiency. A machine, so it was calculated, could excavate in the same number of hours a hundred times as much as a man.

The dredgers were fitted with 'iron buckets fastened to an endless chain, revolving over two drums—one at the end of a long movable arm, to regulate the depth at which the scoops are to dredge, the other at the top of a strong iron framework which rests on the hull of the machine'.[1] They varied in size, according to the work required of them. The largest were of 75 horse power, and cost £20,000 each. They measured 110 feet in length, with 27 feet beam, and drums 48 feet above the water-line. The smallest were of 15 horse power, with an intermediate size between.

In the neighbourhood of Port Said most of the earth thus dredged was used for the reclamation of land and for the manufacture of concrete blocks. When raised from the water the dredgings were 'made to fall into large boxes, having a capacity of 4 cubic yards. Seven of these fit into a barge, which is moored under the spout of the dredge. When all are filled, the barge is floated under a steam crane, by which the boxes are lifted out and placed on trucks running on tramways. On arriving at their destination, one end of these boxes opens on hinges, and the contents are thus readily deposited.' Most of these dredgings

[1] Captain J. Clerk, in *Fortnightly Review,* January 1869.

however were carried out to sea in large sea-going barges, to be discharged into deep water, four or five miles out, through twelve trap-doors, opened and closed by chains, in the bottom of each barge.

For the manufacture of concrete the earth was mixed—in proportions of two-thirds to one-third—with hydraulic lime from France, worked together with water, poured into moulds, and thus solidified into blocks. Nearly 30,000 of these were made, each 12 cubic yards in size and 22 tons in weight. They were left out to harden for three months in the sun, then lifted by cranes into barges and dropped into the sea to form foundations for two long harbour moles—destined to bar the incursions of drifting sand and silt and to enclose a harbour with a uniform depth of 26 feet.

Elsewhere the bulk of the earth dredged from the Canal was discharged into machines specially designed for its disposal. One of these was a *long couloir*, or duct of semi-eliptical shape, 5 feet wide, 2 feet deep, and up to 75 yards long, supported by a tall iron framework which rested on a barge, 96 feet long, fastened to the dredger. These deposited the earth downwards on to the banks. When the dredgings were found to be exceptionally sticky, *balayeurs*—scrapers or sweepers—were used:

This apparatus consists in an endless chain, which is made to pass along the centre of the *couloir*; on this scrapers are fixed at intervals, fitting the shape of the *couloir*. The dredgings are acted on by these much in the same manner as the floats of a paddle act on water. With the assistance thus given, and the current of water, the *longs couloirs* can deliver their dredgings almost on the horizontal line. This application of water-power . . . causes the dredgings to spread themselves over a more extended surface, and in consequence, it settles down firmly, and at a low angle. . . . The float, which supports the *long couloir*, is fastened by chains to the hull of the dredger. By this means the direction of the discharge, as also its distance from the bank, can be readily altered. With the aid of a *long couloir* a dredger can work in the centre of the canal; and by one movement the dredgings are deposited at a considerable distance beyond the water-line, on either side, as may be required.

At a more advanced stage in the excavation, when the dredgers were working at a level farther below that of the embankments, another machine was used. This was an *élévateur*, designed on a similar principle to that of the *couloir* but to carry the earth upwards instead of downwards:

This duct consists of an inclined plane, about 52 yards long, and carrying two lines of tram rail. The inclination is 1 in 4, and it is supported in the middle by an iron frame, which rests on a carriage, running on rails laid for the purpose, along the bank of the Canal, at an elevation of 6 feet above the water-line. The lower end of the *élévateur* reaches over the water, where it is again supported on a steam float. . . . A lighter, containing seven boxes of dredgings, is floated under the lower extremity of the *élévateur*. Each box is raised in succession on to a truck by an endless steel-wire rope, which is adjusted in a few seconds, and it then travels to the upper end of the incline. On reaching this point the box swings vertically, when, by a self-acting contrivance, the door opens and the contents are thus completely emptied.

Of the fruits of this process Lesseps, with his flair for picturesque statistics, declared after the Canal's completion: 'Our dredgers, whose ducts were one-and-a-half times as long as the column in the Place Vendôme, carried off from 2 to 3,000 cubic metres a day; and as we had 60 of them, we succeeded in extracting as much as 2 million cubic metres per month. This is a quantity of which no one can form an exact idea. Let us try to realise it by a process of comparison: 2 million cubic metres would cover the whole of the Place Vendôme and would reach the height of five houses placed one on top of the other; 2 million cubic metres would cover the whole length of the Champs Elysées, between the Arc de Triomphe and the Place de la Concorde, up to the height of the trees . . . It took 4 months to dig out the 400,000 cubic metres of the Trocadero, while we were digging out 2 millions in one single month.' Comparing the Canal works with other large dredging operations, he maintained that in the Clyde it took twenty-one years to dredge three-and-a-half times the amount which the Suez Canal Company extracted in a single month, while at Toulon nine years were required for a similar purpose.

During the first phase of construction, between 1861 and 1864, the total dug, largely by hand, was 15 million cubic metres. During the second phase, between 1865 and the end of 1869, the total dug, largely by machinery, was 60 million. Over roughly eight years, this made a total of 75 million cubic metres. If performed entirely by hand this task, so it has been calculated, would have demanded 25 million man days: 'With an average *corvée* of 20,000 men working 300 days a year, they would have taken twelve-and-a-half years for excavations alone, without taking into account the necessities of port construction etc.'[1]

Now the whole hundred-mile line of the Canal wound off across the desert, marked by the high superstructures of the modern machines, their chimneys curling day and night with the smoke which denoted continuous industry. But it was, in the first place, the industry of the fellaheen which had opened the way across the Isthmus, hollowing out with their hands and their primitive tools the Canal bed which machinery was soon to complete.

*   *   *

But all this was highly expensive. Machine power cost roughly twice as much as man power, even allowing for the fact that its work took roughly half as long. The machines themselves cost some 60 million francs (2,400,000) and consumed coal at the cost of a million francs (£40,000) per month. The six years' delay in completion, beyond the estimated time when the Canal could start to produce revenue, had doubled the interest due to shareholders. For these and other reasons the Canal was costing more than twice as much as its original share capital of 200 million

[1] John Marlowe. *The Making of the Suez Canal.*

francs (£8 million). The Sultan's firman had come just in time to save the Company from a serious and possibly fatal financial crisis, which had been mounting over the past few hazardous years.

Its situation was accurately described at the beginning of 1866 in a report by a British officer of the Royal Engineers, Colonel Laffan, on which Lesseps himself commented: 'Very accurate, and very fair.' Of the Conventions preceding the firman, Laffan wrote that they had proved

> the salvation of M. de Lesseps, for he was literally aground for want of money . . . He had in hand some 45 million fr. in Egyptian Bonds and Shares in his own Canal bought up to keep them afloat on the market. He could not realize these, though he had probably been able to borrow some 30 million fr. upon them with which he enabled his contractors to pay for a portion of their enormous *matériel*. But the contractors would want some 50 million fr. more advanced to pay for the remainder of their working stocks, the total outlay on which would be 80 million fr. And where was Lesseps to find the money? To call up the last 40 million fr. due on the shares would have raised a storm in France, for it would have amounted to a confession of bankruptcy while the real work of the Canal could scarcely be said to have begun. The Pasha's money comes just in time to relieve him from a disagreeable dilemma. It will enable him to defer his call on the shareholders until after the next General Meeting in May next, and it will enable him to work on for nearly three years more.

If, the Colonel added, delays and extraordinary expenses continued, the result might be disastrous to the present shareholders—but not to the Canal itself. For a 'new set' would step in to finish the work. In this case Lesseps, 'entirely absorbed in the prosecution of his great undertaking will take no more heed of the fallen than did Napoleon, when after his retreat from Russia he said to Metternich at Dresden, "What does it matter to me to lose 20,000 men?"'

Concerned though he might be for his shareholders, Lesseps

The breakwater at Port Said

The Canal at Kantara

The waters of the Mediterranean flow into Lake Timsah

was indeed bent, at all costs and by no matter what means, scrupulous or otherwise, on completing his great work of creation. But the cost had now soared, calling for ruthless expedients. That of the contract for the remaining excavations alone, quite apart from the ports, was estimated by Colonel Stanton, the British Consul-General, at more than 150 million francs—a sum equivalent to three-quarters of the Company's initial and now largely spent capital of 200 million francs.

Now, thanks to the firman, the Company was in a position to call up the remainder of this capital, due from the shareholders on the final instalment. But further capital was needed almost at once—a sum estimated at 100 million francs, or half as much again as the original total. Lesseps, indifferent to the mechanics of finance, hence inept in financial discussion, left his associates to worry in Paris while he worked in the Isthmus with his technical experts. And worry they did: it would be hard to raise this money, for the credit of the Company was low and its shares had dropped.

Lesseps thus turned his thoughts once more towards the Viceregal Treasury. In 1866 Ismail had been unable to meet his instalments on the final call. Lesseps had thus raised on his behalf a loan in Paris, apparently without his previous consent, of 17 million francs, at 11½ per cent interest. Of this interest the Company, nicely tided over by such a lump sum, agreed to pay a share. Now, in January 1867, Lesseps proposed to Ismail that he should hand over his entire shareholding in the Company, amounting to more than 85 million francs, in return for the cession of all its remaining lands, installations and buildings in the Isthmus, apart from the Canal itself—a distinctly cynical proposal in view of the fact that the lands had been allowed to it, only a few months before, as being essential to the Canal's completion and maintenance.

The Viceroy was inclined to consider it, partly on the grounds that, in the event of the non-completion of a ship canal, it would give him the land on which to construct a railway between Port Said and Suez, and thus a stronger hold over the Company's activities. But he finally rejected it, heeding the advice of Stanton, who argued that the possession of the shares gave him in fact the stronger hold, which he would be unwise to relinquish, and that their value exceeded that of the property which the Company offered.

Thus the Company resorted to the money-market to raise its required capital of 100 million francs. In an issue of bonds, carrying an interest of 25 per cent, only a third of the sum was subscribed. Once again, the French Government came to the rescue of the Company and its 20,000 shareholders. A law in its interests, based on the advice of the Paris bankers which Lesseps had previously scorned, and on the support of the Empress, on whom he had consistently relied, was introduced into the Assembly by Imperial decree, and passed by both Houses. 'In consideration of the exceptional character of the enterprise and the interest which France has taken in the execution of the Suez Canal', it authorized a special issue of bonds, reimbursable on the analogy of premium bonds, by quarterly lottery drawings, with prizes varying from 2,000 to 150,000 francs, and amounting in the aggregate to a million francs per year. This was designed as an incentive to investment in the remaining two-thirds of the bonds; and within five days the total capital was subscribed by the public.[1] Thus on the very eve of the Canal's completion, the Company was saved from a bankruptcy which its enemies were now gleefully and confidently predicting.

*   *   *

None the less, though the opening of the Canal was now sure, the funds at the Company's disposal remained insufficient. In fixing the bond issue at 100 million francs, Lesseps had under-estimated the actual sum still required. He thus resorted to new and more questionable devices to extract more funds from the Viceroy. He saw a means of doing this through an extreme interpretation of a clause in the original Concession, covering the Company's Customs franchise. This exempted from duty all

---

[1] The same expedient was later proposed by Lesseps in connection with the Panama Canal, but rejected, thus contributing to its financial failure.

imported 'machinery and other material necessary for the work of construction'. The privilege had been stretched and abused through the years, thanks to the tolerance of Said, to cover also furniture, foodstuffs, and consumer goods, not only for consumption but for resale, not only by the Company and its contractors and employees but by merchants and others now established in the Isthmus. As a result Port Said had become in effect a free port and the Isthmus a free zone, open to contraband, which deprived the Egyptian Government of legitimate revenues and moreover exposed it to the hazards of smuggling on a large scale from the Isthmus into other parts of the country. It also threatened an invidious choice between the ruin of Alexandria and the abolition of all Customs dues throughout Egypt.

When, in 1868, the Government established Customs officials at Port Said, it tried at once to put a stop to this abuse, confining the franchise of the Company within legitimate bounds. Lesseps, for his purposes, chose to make an issue of this, insisting on the Company's rights, in view of the 19 per cent of its profits reserved to the Egyptian Government, to import free of duty anything not consigned to other parts of Egypt. He also drafted a circular, encouraging the Consular Corps to do the same, in vindication of the rights of its nationals. He hinted that he was prepared to provoke a conflict with the Egyptian Government, with a view to a further Imperial Arbitration. In this course he did not have the support of the French Consul-General, Poujade. The Consul sympathized with the view of Sherif Pasha that the Egyptian Government could not thus tolerate the creation of a State within a State, which would make of the Isthmus a nest of smugglers. In August 1868 Lesseps forced a showdown, encouraging four French merchants to remove a consignment of goods from the Customs by force and without payment of duty.

Lesseps then proceeded to Constantinople, where the Viceroy was pursuing his own cause at the Court of the Sultan. Ismail appointed a Commission, which put forward an equitable solution, extending the Company's agreed franchise to cover food, drink, clothes and medicines, but denying its right of delegation to merchants and others. After a protest this compromise was accepted by Lesseps, whose claim was in fact little more than a subterfuge. His real motive was to secure an increased Customs concession, not for its own sake but for the sake of renouncing it and claiming compensation from the Viceroy for doing so.

This he now did, putting forward also a series of further proposals. He proposed to renounce against payment various other specified rights which the Company enjoyed, and all other outstanding claims against the Egyptian Government. He reverted, in a new form, to his original proposal regarding the lands along the line of the maritime canal, still held by the Company, now proposing—with a bland disregard of the Imperial injunction that they should not become a source of profit—that they should be sold in plots to the joint profit of Government and Company. Finally, he proposed that installations not now required for the operation of the Canal, for example telegraph offices and hospitals, should be taken over by the Egyptian Government for its own administrative needs. Such were the shifts to which Lesseps and his Company were reduced in order to pay for the completion, on time, of the last lap of the Suez Canal.

Ismail agreed to this one-sided bargain for his own personal motives. He was still pursuing his cherished ambition of increased independence from the suzerainty of the Porte. By a firman in 1867, he had achieved the long-coveted right of primogeniture, in his line of succession, together with the title of Khedive. But he still coveted the further right to negotiate his own treaties with foreign powers; and he believed that he could further these aims through the support of Lesseps and the Suez Canal Company. This was to prove an erroneous belief. The Khedive Ismail was in fact left to fight his battle with the Porte alone, receiving only qualified support from France or Britain. Neither of them anxious to see Egypt severed politically from the Ottoman Empire, they were mistrustful of Ismail's increasing financial extravagance and military ambitions. It was a battle to be won only four years later, under a new Grand Vizier.

Meanwhile the Khedive thought it worth his while to satisfy the Company's need for funds. On April 23, 1869, a Convention was agreed, by which he would pay 20 million francs to the

Company for the privileges ceded, and a further 10 million for the installations. In a second Convention he agreed also to the joint exploitation of the lands, subject to joint agreement on a revision of the present system of Capitulations. This would regulate, through the establishment of Mixed Courts, the legal position of foreigners in Egypt, in relation to one another and to Egyptian subjects.

These 30 millions were raised by the Viceroy's renunciation of interest and profits on his shares in the Company for twenty-five years. This was done through the detachment, from the shares, of coupons, and their offer in the form of 120,000 bonds, bearing 5 per cent interest and dividends over this period, to other Ordinary shareholders. Here was a handsome bonus for the shareholders, who had lately been through anxious times; and a handsome bonus for the Company, which the Viceroy had in effect indemnified twice over for concessions twice repeated, and for installations which had served their purpose and which it no longer needed to use. Two years later, by a provision depriving of voting rights the holders of shares from which coupons had been detached, it denied to Ismail all influence over the Company, whose principal shareholder he was. The Viceregal barrel had been scraped to the dregs.

By the end of 1869, the Viceroy of Egypt had contributed to the cost of the Suez Canal a sum amounting to some 240 million francs (just short of £10 million). His contribution consisted of 87 million francs in share capital, 124 million francs in indemnities —in effect the recovery of lands and rights which had originally belonged to him—and 30 million francs in interest on arrears on these sums. It amounted to more than half the total cost of construction—450 million francs (£18 million)—a sum more than double the original estimate. It took no account of the cost of public works indirectly concerned with the Canal—mainly the construction of the harbour at Suez and of the westerly stretch of the Sweet Water Canal—which amounted to some 55 million francs; nor of the incidental expenses of Nubar's missions abroad or of the lavish festivities staged for the opening of the Canal. In all, Ismail's final Canal debt amounted to some 400 million francs

(£16 million), one-fifth of that final total debt, in foreign loans, which was to lead him to bankruptcy seven years after it was opened to traffic. Such was the cost to Egypt of a Canal from which only Egyptian posterity would benefit.

*   *   *

From now onwards the construction proceeded with several alarms but no crucial mishaps, working to a completion date postponed from 1868 to October 1869, with bonus and penalty clauses agreed in the contract. One after the other the physical obstacles were removed from its path. The cutting through the ridge of soft sand at el Guisr, between Port Said and Timsah, was at last completed in January 1868, six months ahead of contract, after more than eight years of excavation, primarily manual but finally mechanical. The cutting through that of the Serapeum, between Timsah and the Bitter Lakes, could be manually excavated only down to a certain point. Then a thick vein of gypsum was found, and the dredgers took over, operating with the aid of explosive charges in a bed temporarily flooded by the Sweet Water Canal. This work was soon completed so far as to allow the maritime Canal to be prolonged to the northern shore of the Great Bitter Lake, where a 300 foot barrage contained it.

Here, in the middle of March 1869, a striking ceremony was performed by the Khedive Ismail, who chose this moment to pay his first visit to the works in the Isthmus. The waters of the Mediterranean were let loose for the first time through the sluices of the barrage, to cascade in a huge wave of foam into the Bitter Lake—'that yawning basin', as a spectator described it, 'stretching as far as the eye could see', where soon 'a great sea would be created'. Lesseps, suiting his words to the solemn occasion, said to Ismail: 'Moses ordered the waters of the sea to retire, and they obeyed him; at your orders, they now return to their former bed.' Ismail was impressed by the Canal between Port Said and Kantara, which he saw as 'a veritable Bosporus'. But he voiced his misgivings as to whether the rest of it—especially the stretch from Ismailia to Suez—could be made navigable on time. To finish it, he estimated, would cost 10 million francs. Lesseps had timed the Khedive's visit well. It was a month later that he signed the Convention giving the Company the final millions it needed.

Close upon the heels of the Khedive came the Prince and Princess of Wales, on a formal visit to the Isthmus. Arriving at Suez they were entertained to dinner at the hotel, 'well-known to Indian voyagers' and served by turbaned Bengali servants with a dinner containing 'several excellent Indian dishes'. Next morning, after inspecting the harbour of Suez and the mouth of the Canal, they proceeded by train to Ismailia, stopping on the way at Chalouf. Here Lesseps presented the Princess with a bough of orange blossom grown in the Company's garden, and the party proceeded by horse and carriage to the site of the works. Here, where no water had yet been introduced, they surveyed a cutting as wide and deep as a valley, descending in a series of inclined and graded planes to a network of tramways below. Thousands of workmen were loading the soil into trucks, which were drawn by mules to the foot of the incline and thence raised by a long rope, worked by an engine, to the head of it.

Farther north, at the dam which the Khedive had first breached, the Prince opened a further series of floodgates, and more water rushed through into the neck of the lake, where Mediterranean fish were already disporting themselves in a pool. Less happily, where the waters of the Sweet Water Canal had been emptied a few days before into those of the maritime canal, 'millions of fresh-water fish, on which crowds of gulls are now feeding, met their death in the water coming from the salt sea'. To *The Times* correspondent, surveying the huge extent of lake to be flooded, 'it really seemed as if we were trying to fill the London Docks out of a bottle';[1] and indeed the process was expected to take five or six months.

To Lesseps the Prince of Wales confessed his regret that the British Government had allowed French enterprise to accomplish this work, so vital to communication with India, and he criticized Lord Palmerston's lamentable lack of foresight in refusing to give it support. Six years later, passing through the completed Canal

[1] December 7, 1869.

on his way to India, the Prince repeated his views to Lord Granville. 'The Suez Canal', he wrote, with a prophetic touch, 'is certainly an astounding work, and it is an everlasting pity that it was not made by an English company and kept in our hands, because, as it is our highway to India, we should be obliged to take it—and by force of arms if necessary.'

The basin of the Bitter Lakes, almost alone along the line of the Canal, presented few if any constructional problems. More than 25 feet below sea level, it would provide, when both seas had filled it, a navigable channel of ample depth. At the bottom of it was a long thick bank of encrusted salt, probably formed by evaporations of salt water as the Red Sea receded. But this dissolved easily enough when the waters returned. Those of the Mediterranean were now gradually filling the lakes; but it was not yet possible to add those of the Red Sea. For another difficult obstacle remained to be conquered between the lakes and Suez— the ridge of Chalouf, whose deep dry cutting had impressed the Prince of Wales.

Sir John Hawkshaw, in his report, had diagnosed the existence of a stratum of rock in this ridge. But Lesseps, with his airy optimism, had brushed such fears aside, admitting only that a ledge of rock had indeed been found and that his engineers had made a short curve to circumvent it. None the less, at a level above the minimum required depth of eighteen feet, a mass of conglomerate rock was struck, which had escaped notice earlier. It would have been a task of some magnitude to remove this by dredging, had there been water in the cutting above it. Fortunately it was still dry, and the rock was removed by a long process of blasting, while manual labour carried the débris away. In the course of this, fossil remains were found, including those of a number of sharks.

All along the line an intensified effort was being made to ensure the completion of work on time: 'Thousands of men were employed—Dalmatians, Greeks, Croats, negroes from Nubia, and Egyptian fellahs, all superintended by French officers. These gangs of men were regularly organized and paid according to the cubic feet of earth they dug out, some earning five or six, and others only two or three francs a day. The works were pushed on with great rapidity, steam-traction and railways, asses, mules, men and camels, all contributing towards their completion.'[1]

Rumour ran rife, causing undercurrents here of complacency and there of concern, that the Suez Canal would not be completed in time for the inaugural ceremonies. By the middle of July 1869, $6\frac{1}{2}$ million cubic metres of earth, out of a total of 75 millions, still remained to be extracted, and indeed this task was not completed until after the opening. But on August 15, in a preliminary ceremony, the waters of the Red Sea met and mingled with the waters of the Mediterranean, doubling the rate at which the lakes would be filled and establishing them as a buffer to neutralize any slight difference of level between the two seas. Early in October a French ocean-going steamer made the journey from one sea to the other.

But there was a last-minute crisis. A mere fortnight before the opening date, a large kidney-shaped ridge of rock, too hard to be worked by the dredgers, was unexpectedly struck at the Serapeum. It projected so far above the bed of the Canal as to leave only ten feet of water. Never, until this emergency, had blasting been tried under water. The experts were pessimistic: the Canal, they feared, could not now open on time. But Lesseps exclaimed to them: 'Get gunpowder from Cairo, masses of gunpowder, and if we can't blow up the rock we shall blow up ourselves.' The channel was cleared, with twenty-four hours to spare, down to a depth of sixteen feet instead of the required twenty-six, and for the opening convoy ships were limited to those drawing less.

Thus the Suez Canal, a hundred miles long, eventually twenty-six feet deep, seventy-two feet wide at the bottom but some two to three hundred feet wide at the top, was ready to be thrown open to the steamships of the world. Man had irrevocably joined the two seas—'the sea of pearls and the sea of coral', as Lesseps quoted, fulfilling an ancient Oriental legend, and possibly re-establishing a union which dated back to a remote geological age.

[1] Captain Clerk in *Fortnightly Review*.

CHAPTER XV

The Opening Ceremonies

The Khedive Ismail shrewdly judged that the opening of the Suez Canal should give Egypt a new status in the eyes of the powers of Europe. It must be exploited to add lustre and prestige to his name and his dynasty, moreover to assist his ambitions for further independence, at the expense of the Porte. Ismail thus determined, regardless of cost, to dazzle the world with a sumptuous programme of inaugural ceremonies. In July 1869 he set forth on a tour of the capitals of Europe, personally inviting its crowned heads to the opening, zealous as he travelled to cut the figure of a reigning sovereign, and aspiring to enlist their ultimate support for his cause. This gesture of independence displeased the Porte, which was neither consulted nor informed in advance of his journey.

The invitations, limited only by the fact that he possessed a mere eight palaces in which to house royal guests, brought various replies. The King of Prussia regretted his inability to accept, on account of his age and the fatigue of such a journey, but he would send the Crown Prince instead, as his official representative. The Emperor of Austria had planned likewise to send an Archduke, of the Imperial family, but, rather than thus be outdone in precedence by Prussia, decided to come in person, visiting the Sultan in Constantinople *en route*. The Kings of Greece and of Sweden and Norway refused, rather than give offence to the Sultan. The King of the Netherlands would send his brother, Prince Henry. Russia would be represented by her Ambassador at the Porte, General Ignatieff. The President of the United States, General Grant, was not in a position to accept the invitation in person, nor could he send an official representative since Congress, whose right it was to appoint one, would not meanwhile be in session.

The Duke of Aosta would represent the King of Italy, accompanying an Italian naval squadron, but was in fact obliged to leave with it before the inauguration. The Khedive resisted, for lack of worthy accommodation, the desire to invite the Sultan of Morocco, the Shah of Persia and the Bey of Tunis; also the Kings of Bavaria, Württemberg, Saxony, and the Grand Duke of Baden, whose invitations would in any event have required the tacit approval of Prussia. Even Egyptian hospitality perforce had its limits.

The Khedivial tour culminated in London. Ismail's previous visit in 1867 had been awkwardly timed to coincide with that of the Sultan, who inevitably overshadowed his Viceroy. This time, on the day of his arrival, *The Times*[1] published a leading article, referring to Ismail's recent princely hospitality to the Prince of Wales in Egypt, and urging that he should be as liberally entertained, with no 'stinted ceremonial' in return. It disparaged the too 'cautious nicety' of the Government's courtesies, over-responsive to the Porte's insistence that the Khedive was not a reigning Prince but a vassal of the Sultan. Next day indeed the *Pall Mall Gazette*[2] ran an article entitled 'How Not to Honour Him', revealing that Ismail on arrival had been shown into Buckingham Palace, where he was staying, not by the main entrance but by a side door, and retailing (incredulously) 'that diplomatic representations have been made to the effect that the Viceroy must not be allowed to sleep in the bed occupied by the Sultan'.

Honour, however—and *The Times*—was satisfied by his subsequent entertainments. Ismail stayed not merely at Buckingham Palace, in the Queen's absence, but at Windsor Castle, in her presence, and was entertained by the Prince of Wales to a sumptuous festival in his honour at the Crystal Palace, culminating in a display of fireworks whose 'novel, beautiful, pyrotechnic effects' included a set-piece with three resplendent crescents in a circle of brilliants.

Ismail hoped that the Queen would designate the Prince of Wales to represent her at the inaugural ceremonies, and afterwards

[1] June 23, 1869.   [2] June 24, 1869.

wrote to her accordingly, stressing that the opening of the Canal would bring together her possessions in Europe with those in the Indian Ocean. But Her Majesty's Government decided otherwise. The Prince of Wales was obliged to refuse the invitation, and the Queen would be represented instead by Mr. Henry Elliot, now British Ambassador at the Porte. In any event the star guest at the ceremonies was to be the Empress Eugénie, whose benevolent influence had smoothed the path of the Canal's construction, and who was now officially to declare it open.

Crowned heads apart, the Khedive invited some thousand guests—guests in the grand manner, with expenses paid, including transport from Europe to Egypt and back again, and in the country itself board and lodging, entertainments, carriages, horses, passes on the railways, and no call to put a hand in the pocket for even the hire of a donkey. All *baksheesh* was covered by the Khedive, and extra personal expenses readily refunded, on request, by the Master of Ceremonies.

As an advance guard, a select hundred celebrities were treated to a three-week trip up the Nile to Assuan and down again, before the inaugural ceremonies, exploring the monuments of the Pharaohs in the company of Egyptological experts. All the talents were represented among them, not only of the official but of the professsional world: the world of industry and banking and trade; the world of learning and science; the world of the arts; the world of sport and of the Press. Here in strange juxtaposition, gazing at the unfamiliar sights of the Orient, was a cross-section of courtiers, diplomats, deputies, soldiers, lawyers, doctors, technicians, chemists, anthropologists, physiologists, naturalists, members of the Jockey Club, painters, sculptors, dramatists, men of letters, historians, and journalists. Of the more distinguished, the majority came from France, but all countries were represented, with Ibsen from Norway, and Switzerland, on Ismail's explicit instructions, not forgotten, for 'by her commerce and her industry she is called upon to have frequent relations with Egypt'.

This privileged selection of guests, which would hardly have shamed Napoleon's *Institut d'Egypte*, reached Alexandria in two ships from Marseilles, on October 15, 1869. They were met by

Lesseps in person, and proceeded to Cairo by train. Next to Zola and Dumas, the most notable writer among them was Théophile Gautier, who observed his fellow-travellers with amusement, and was especially intrigued by the variety of hats which they wore. They had landed, in a temperature only a few degrees warmer than that of Marseilles, already equipped to resist the extreme heat of Upper Egypt. Many were thus wearing white sun-helmets, familiar enough throughout British India as *solar topees*—and indeed, doubtless recommended by Thomas Cook, who had lately set up his travel agency and was present as a personal guest of Lesseps—but bizarre to the eyes of a Frenchman fresh from Paris. Gautier likened them, with their green shaded 'visors' and neck-protectors, as though fashioned from mail, to the helmets of the ancient Saracens.

Others wore blue veils rolled like turbans around their helmets. Others again, less versed in local colour, wore felt hats, or panamas lined with green taffeta, while one charming old *savant,* renowned in the world of chemistry, wore a stove-pipe hat, black coat, white tie and unlaced shoes, insisting that he was used to this costume and would feel naked if otherwise dressed. Gautier observed also, shading the eyes of the visitors, a large variety of spectacles, some of blue glass, some of glass smoked as though for viewing an eclipse, some with blinkers in protection against all kinds of weather, and in particular against the ophthalmia which was known to rage in Egypt.

Arriving in Cairo the guests were distributed, according to their varying degrees of distinction, among hotels of varying quality. Among them was a sadly superannuated literary light, Louise Colet. In her time a stormy Romantic poetess, muse of Victor Hugo, confidante of Madame Récamier, and mistress of Flaubert, she was now disgruntled to find herself lodged in an inferior hotel and consigned, on the journey up the Nile, to a rat-ridden boat and the most obstreperous donkeys. On one excursion, too tired to move a step farther, she was placed in a sack and carried back to the boat by a team of fellaheen, to the keen amusement of all.

Théophile Gautier found himself installed at Shepheard's, a

hotel with which Cook was associated. It had been partially destroyed by fire earlier in the year, but hurriedly rebuilt and enlarged, and now resembled, to French eyes, an English barracks rather than an Eastern caravanserai. To Charles Blanc, another French writer, it was 'a vast convent, whose staircases and corridors are hardly lit at all, and whose enclosed rooms resemble monastic cells. One understands here that light and heat are the same and that coolness is synonymous with night.' He took pleasure in walking through the silent sombre corridors, imagining himself in a monastery. From time to time a ray of light shone suddenly—through the door of a cell which a monk had opened. He ran into one of his own party, dragging himself along with his arm in a sling and a companion supporting him. It was Gautier, who explained in his slow asthmatic voice: 'You see what comes of embarking on a Friday . . . I opened the door of a room which I took to be that of a friend: I fell into a coal hole and have broken my collar bone. So I am condemned to remain here.'

Gautier thus spent most of his visit watching the world go by, with a none the less observant gaze, from the terrace of Shepheard's. Of its restaurant he remarked on the

excellent French cuisine, lightly Anglicized, natural enough in a house whose normal clientele is almost entirely English. No Arab dishes were served by a slave of dusky hue, in white turban and rose-coloured robe . . . but we did not regret it too much, local colour being often more agreeable to the eye than to the palate. The travellers were grouped at table according to their chosen or professional affinities: there was the corner for painters, the corner for *savants*, the corner for men of letters and reporters, the corner for men of the world and connoisseurs; but all without strict delineation. One paid visits from one corner to another, and over coffee, which some took Turkish and others European, conversation and cigars blended all ranks and countries; one saw German doctors talking aesthetics with French artists and grave mathematicians listening with smiles to journalists' gossip.

On October 18 the guests were received in audience by the Khedive. He addressed them as 'an accomplished agriculturist and industrialist', well aware of the value of his country's resources, and supporting his statement with figures. The fellah, he assured them, was not unhappy. 'He reaps from his patrimony more than could be reaped elsewhere. His troubles are lessened by the fertility of his soil. The Nile is an excellent worker.' That evening he gave a gala reception at the Kasr el Nil Palace, at which Madame Colet, dreaming of *Arabian Nights* entertainments, was disillusioned by a 'counterfeit of the Tuileries', with frock-coated guests and a colourless performance of a *Caprice* by Alfred de Musset.

Meanwhile Lesseps, familiar to all as Monsieur le Comte, kept open house in the Hotel d'Orient, which he had made his headquarters. Modest in comfort but French in fare, it overlooked the Ezbekieh Gardens, once a lake overlooked by the palace in which Napoleon had lodged during his brief occupation of Cairo. It lay on the fringe of the medieval city, so rich in mosques and monuments. These the visitors duly appreciated, regretting only that their façades had been painted up for the occasion with a thick red and white wash, designed to brighten them but in fact obscuring much of their fine Arab stonework and ornamentation.

*   *   *

On October 22 the Empress Eugénie and her suite reached Alexandria in the Imperial yacht, the *Aigle*, from Constantinople. *En route* she had paid unofficial visits to Venice and Athens, but her visit to the Sultan was official, and inspired by diplomatic motives. She hoped to soothe his irritation at the fact that the Khedive had invited Heads of State to the inaugural ceremonies without so much as informing him that any were planned.

The Empress was handsomely received by the Sultan, and lodged in the Beylerbey Palace on the Asiatic side of the Bosporus, with the luxurious imperial barge at her personal disposal. To the Grand Vizier she explained that she herself, at the Paris Exhibition of 1867, had expressed a desire to attend the opening, and that the Khedive thus had no choice but to invite her. On receiving her invitation she had enquired whether the Sultan too was invited. To this the Khedive replied that he could not without impertinence invite his sovereign to a house which was the sovereign's own. In fact, as the wary Ismail knew, the Sultan could have announced his arrival self-invited, and stolen the thunder by acting as host to his Viceroy and guests. Right up to the eleventh hour the Sultan's Ministers sought to persuade him to do so—but without success.

Now the Khedive met the Empress at Alexandria and conducted her to Cairo by train, lunching at a station *en route*. A French reception awaited her in Cairo, where she drove to the Gezireh Palace beneath an *arc de triomphe* erected in front of the Consulate. But since this first stage of her visit was unofficial she was still free from formal engagements, free to absorb the atmosphere of Cairo, under the guidance of Ismail and others, at will. She soon felt at home, for as she wrote to the Emperor, the country reminded her strongly of Spain. The music, the dancing, the food seemed identical. The dances in the harem resembled those of the Spanish gipsies, but were 'perhaps more *indecent*'.

Much of her time was devoted to sight-seeing. Her guide was the distinguished French Egyptologist, Mariette Pasha, a huge man, forbidding but courteous, in black spectacles and a red *tarbush*, who had been launched on his career with the help of Lesseps, building the Cairo Museum and establishing an Egyptian Department of Antiquities. In the course of a tour of Upper Egypt, he found the Empress a willing and well-informed pupil, since she had taken a brief preparatory course in Egyptology in Paris under a young professor named Maspero.

A visit to the Pyramids was postponed until her return to Cairo, since Ismail had ordered a seven-mile carriage road to be built for her, from the city to Giza, and it was not yet ready. Pending its completion no foreigners were permitted to visit the Pyramids, and an Englishman, who obstinately insisted on doing so, soon saw the reason why. For the road was being built by intensive forced labour, spurred on by an unusually free use of the *kurbash*.[1]

---

[1] Edward Dicey. *The Story of the Khedivate.*

The opening of the Canal: religious ceremony at Port Said

The first convoy steams through the Canal

The Empress Eugénie with Lesseps at Ismailia

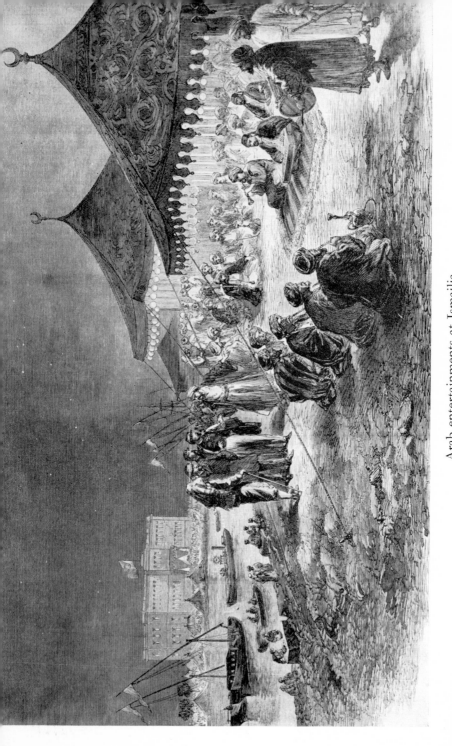

Arab entertainments at Ismailia

Meanwhile the Khedive had hit upon the idea of commissioning Mariette to write him the libretto of an opera on an Egyptian theme. It was to be set to music by Verdi, and performed for the inauguration ceremonies in the Opera House which he was starting to build. The archaeologist, who had written poetry and other literary works in his youth, was delighted at the prospect, and wrote the story of *Aida,* placing it within the framework of his recent archaeological researches. He broke the news of his commission in a letter to his brother in Paris, adding the words: 'The Viceroy is spending a million. Don't laugh; it is perfectly serious.'

Unfortunately Verdi at first turned down the proposal, which in any case he could hardly have executed in time for the occasion: it took longer to turn out an opera than to run up an opera house, as Ismail now did, from lath and plaster in an elegant rococo style. On reading Mariette's synopsis of his story, Verdi changed his mind in the following year, and *Aida* was first performed in Cairo in 1871. Meanwhile the Opera House was opened, early in November 1869, with a performance of *Rigoletto,* preceded by a solemn cantata on the Khedive's social and economic achievements for his country, embellished by a *corps de ballet* of forty young Italian girls, whose virtue was protected behind bars in the neighbouring police station. The opera was only momentarily marred, at the second performance, by a fire on the stage, which drove the actors to jump into the orchestra pit until it was extinguished, while Ismail, from his box, calmed the audience. This launched a season of Italian opera, which added to Cairo's gaieties throughout the winter season.

By now the Khedive's nine hundred remaining guests had begun to swarm into Egypt, 'a self-inflicted plague' as *The Times* described them,[1] over-crowding the hotels, and placing much strain on the organization, formed and superintended by Nubar to lodge, feed and entertain them. Usually they slept two or more to a room; 'noble lords . . . shoved into two-pair backs and doubled up with "commercials", or literary gents, or clods'. One dissatisfied guest demanded a personal audience with the Khedive,

[1] December 7, 1869.

to whom he complained that he had been put to sleep with another in a double-bedded room. Ismail atoned 'with a fine stroke of humour', ordering that three more beds should be placed in it. For the supply of food, wine, waiters, linen, plate and tents, the Egyptian Government had invited the leading caterers of the Levant to submit tenders, on the understanding that the contract would go to the lowest. But, according to popular rumour, hints were given that it would go in fact to the competitor who offered Ismail himself the largest commission—a business deal robbing Peter to pay Paul, since the Treasury of the Khedive and his Government were one and the same.

Most of the newcomers proceeded direct, in slow and congested trains, to Ismailia, which was to be the centre of the inaugural ceremonies. Here the Khedive had to cater in all for some six thousand guests. The Europeans were accommodated for the most part in long lines of tents, well-furnished and carpeted, with a marquee, containing a huge permanent buffet, to each canton-ment, while the Canal engineers and officials gave up houses and gardens for conversion into dormitories. To help feed and serve them Ismail imported five hundred cooks and a thousand servants from such places as Trieste, Genoa, Leghorn and Marseilles. Each day they consumed whole flocks of sheep, which arrived each morning to be killed and roasted in the evening, in 'gargantuan and pantagruelian kitchens'. Meanwhile a ship had arrived from Bordeaux with 'a whole cargo of excellent wines', so that soon the shores of Lake Timsah were 'littered with empty bottles'.

Entertainments for the visitors, in marquees, abounded—in some coffee to the strains of native music; in others 'exhibitions of jugglers; in others religious exercises—dancing, or swaying, or singing, chanting, grunting dervishes; in others snake charmers, serpent swallowers, glass-eaters; in others reciters, story-tellers; in some singing-women concealed behind their curtains of gauze. . . . At night all these tents are lighted with lanterns, and the streets and passages between the tents are lined with bon-fires and with iron frames on poles filled with blazing pine.'

Amidst all this the Arabs, well-behaved and courteous, 'sit packed in rows on their hams—white turbaned, dark robed, grave, fine-looking men. It would take the conceit out of the grand Europeans, who fancy they are the flower of the human race, to walk through a crowd of these splendid children of the Desert, and measure their stature and breadth of shoulder against those of the man whose drink is water and whose food is millet, maize and vegetables.'

Hundreds of Arab chieftains and dignitaries had come with their tribes and their herds from all parts of the Islamic world, their encampment becoming a huge native 'town' in the desert. One aged Sheikh, asked why he had brought his tribe so far, could only say he believed that the French girl (in Arabic *bint*) wanted to see them. Ismailia, as it awaited the ships, was to Charles Blanc a Babel of tongues, a 'macedoine of races, yellow, black, or copper-coloured, a mob of camels, donkeys, horsemen and carriages, a Fair of St. Cloud multiplied by thirty thousand Arabs and transported to beneath the skies of Egypt'. To the Englishman from *The Times*, it was 'Greenwich Fair, and Bartlemey Fair, and Donnybrook *à l'Arabe*—that is, without women'.

$$* \quad * \quad *$$

Meanwhile, in the harbour of Port Said, a large concourse of ships was assembling, ready to pass, on the appointed date, from sea to sea. The first royal guests to arrive were the Prince and Princess of the Netherlands, in their yacht. Next came Francis-Joseph, the Emperor of Austria, in his own, with a naval frigate to escort him. Visiting Jerusalem *en route*, he had, in his determination to arrive on time, made a hazardous embarkation in a storm from an open beach at Jaffa. He was closely followed by another royal yacht, containing the Crown Prince of Prussia.

Finally a convoy of twenty ships appeared on the horizon, to be signalized by salvoes of artillery fire from the shore batteries and the dressed ships at anchor. The *Aigle*, with the Empress Eugénie on board, sailed amid a tumult of cheers into a harbour already dense with the be-flagged ships of all nations. Numbering more than eighty, of which fifty were warships, they included, apart from the royal yachts and escorting naval vessels, a British Iron

Clad Squadron, a Russian corvette, and frigates from Spain, Denmark, Norway and Sweden. When the French imperial yacht had anchored in the midst of them, the Khedive, with Ferdinand de Lesseps and his two sons, hastened on board to pay his respects to the Empress, to be followed later in the morning by the Emperor of Austria and the other royal Princes. It was, as she remarked, 'a magical reception'.

In the afternoon, on the beach before the Quai Eugénie, the Canal was blessed before her in a double religious service, Moslem on the one hand, Christian on the other. No such service, according to Lesseps, had ever before been held in any oriental country. With the approaching union of the two seas, it sought to symbolize that of the two Faiths—the union of man before God. Lesseps awaited the Empress and the Khedive at the stand which had been erected for the royal and official congregation.

On a platform to the left, still to the accompaniment of gunfire, the Grand Ulema read aloud a simple discourse, following the Moslem prayer. On a similar platform to the right, the mitred Bishop of Alexandria officiated in a Christian *Te Deum*. This was followed by an eloquent sermon, preached by Monsignor Bauer, Almoner at the Tuileries and Protonotary Apostolic. Waiting for the gunfire to subside, he launched into high flights of rhetoric, lauding those whose Faith and whose superhuman efforts had, by this grand work of civilization, united the Eastern world with the Western. He apostrophized in particular this man of genius, Ferdinand de Lesseps, comparing his achievement to that of Christopher Columbus, and consigning him to a history into which, by a rare dispensation of Providence, he was now entering in the course of his lifetime.

That evening the Empress, incognito, walked through the streets of Port Said by the light of the fireworks. Among the visitors some found it a charming little town, which recalled Arcachon, in the Landes; for others it evoked the improvised towns of Australia and America, one comparing its peculiar blend of business and pleasure, of restless speculation and crude entertainment, to that of a miniature San Francisco. But all agreed that here was a first-class port, with a great commercial future ahead of it. Already, before the official opening, it was coming to be used as such, its ships conveying not merely coal to Suez but pilgrims *en route* to Mecca, and exporting cotton and cereals to the other ports of the Mediterranean.

Then the historic day dawned on which the Suez Canal was to be put to the ultimate test. The date was November 17, 1869, Lesseps's sixty-fourth birthday. In the course of the previous night there was a last-minute mishap. An Egyptian corvette, the *Latif*, had been sent ahead with a French ship to try out the channel. Before reaching Kantara her captain made a false manœuvre; she strayed from her well-marked course and ran aground on the bank, protruding into the canal. A salvage boat, sent from Port Said, failed to dislodge her. At three o'clock in the morning the Khedive himself, on his way to Ismailia in a naval frigate to receive his guests on arrival, turned back, and met Lesseps at the spot.

Either, they agreed, the *Latif* must be towed back into the channel, or she must be towed into the bank. If both these courses failed there was, hinted Lesseps, a third alternative. Ismail looked him in the eyes and exclaimed: 'Blow her up! Yes, yes, that's it. It will be splendid.' Lesseps embraced him. The Khedive smiled: 'But at least', he added, 'wait until I can remove my frigate and let you know that the coast is clear.' Rumours were to reach Ismailia that he had threatened to impale some of the offending officers. All such drastic steps proved unnecessary. With the aid of the Khedive's substantial crew, the *Latif* was refloated and reached Kantara, where she anchored and awaited the convoy's passage, dressed with flags and preparing to fire a royal salute.

The ships of the convoy, forty-six in number and divided into five divisions, each a thousand metres apart, left at five or ten minute intervals from eight-thirty in the morning onwards. The *Aigle* sailed at the head of it, entering the Canal between two colossal pyramids of wood, fashioned for the occasion. The royal and ambassadorial yachts, with their escorting warships, followed. The Russian ship failed to get into her allotted place in the line, thus losing precedence to the *Psyche,* with the British Ambassador on board, which went ahead of her. The *Péluse,* pride of the local

French mercantile marine, with the officials and guests of the Company aboard, brought up the rear of the first division—and, from its size, was to run once or twice into trouble.

The Empress, fearing that some mishap to the *Aigle,* however slight, might halt the ships and so compromise the honour of France and her flag, was nervous throughout the voyage. She asked for Lesseps, who was not at first at her side—and he was found prone in his cabin, sleeping off like a child the anxieties and efforts of the night before. At one moment, unable to disguise her emotion, she retired to her own cabin, and burst into tears. But all went well, and soon the convoy was passing smoothly through the cutting of el Guisr.

Pressed together on its ridge and strung out along the banks, was an immense crowd from Ismailia. At five-thirty, as Fromentin (who had come to Egypt with his notebook, not to paint but to write) described the scene:

> a light coil of smoke and the tip of a high mast appears above the high sandbanks of the Northern Canal. From one mast of the ship, still hidden, flies the Imperial flag of France. She is the *Aigle.* She passes beneath us slowly, her wheels barely turning, with a cautious prudence which adds to the solemnity of the moment. Finally she emerges into the lake. Salvoes of artillery from all the batteries salute her, the immense crowd applauds, it is truly wonderful. The Empress, from the high poop, waves her handkerchief. She has M. de Lesseps at her side; she forgets to shake his hand before this great multitude, come from all parts of Europe and overcome with emotion.

For a moment the crowd stood in breathless silence. Then it burst into cheers, throwing hats in the air and embracing, crying: 'Long live the Empress! Long live Lesseps!' Entering the waters of Timsah, the *Aigle* was saluted by a French and three Egyptian warships, which had sailed up from Suez. As soon as she was at anchor the Khedive hurried on board, paid his respects to the Empress and hugged Lesseps in a warm embrace.

That evening Lesseps presided over a banquet to the members of an International Commercial Congress of the European

Chambers of Commerce, brought together for the inauguration. In this influential international company, he shrewdly forsook rhetoric and conventional phrases of congratulation at what had been done. He stressed instead, in practical and critical terms, what remained to be done. He enlarged on the outstanding legal problem, that of the reform of the old system of Consular tribunals in Egypt, based on Capitulatory rights ceded to each power by the Sultan. This was now seriously impeding the work of the Company and the country's commercial development. A joint Commission, under the terms of the final Conventions, was now considering the problems. Lesseps urged his audience to put pressure on its Governments to modify their several rights and formulate a new joint system. The Egyptian Government was willing; but the French Government, he frankly stated, was proving obstructive. The Canal Company had thus drawn up a petition, and for this he begged international support.

Next day Ismailia was *en fête*. In the morning the Empress rode with Lesseps to el Guisr, where she looked down on the Canal from above. She wore a white riding-habit, and a light helmet made of elder-wood, covered in white cloth, which suited her better than it suited her ladies-in-waiting. She insisted on riding back into the town on a dromedary. Fromentin saw her pass. 'She rides well', he noted. 'She looks tired.' She proceeded to Lesseps's house, where she received the ladies of the Isthmus. They admired the large silver cup which she had given to Lesseps, with the Grand Cross of the Legion of Honour. He was to receive similarly high Orders from the Khedive and the Emperor of Austria. But he refused the Emperor Napoleon's offer to dub him Duke of Suez.

That afternoon there was an equestrian *fantasia*, with displays by galloping Egyptian cavalry and Beduin horsemen, firing in the air as they rode, and a six-mile race of dromedaries, 'with screaming Arabs on their humps'. Arab musicians performed on flutes and tambourines and big drums, while European ladies drove in their carriages through the stifling crowds and the shifting sands, upright on cushions in *grande toilette* as though attending the races at Longchamps. A correspondent of *Blackwood's*

Magazine[1] was not impressed at the display, nor at the turn of speed of the Arab riders. It did not seem to him 'at all beyond the achievements of an English dragoon or good rider to hounds'.

In the evening the Khedive gave a great ball for the Empress in the palace which he had hurriedly built in the Moorish style. Its upper floor was not yet finished, and its gilded rooms, vast as they were, were too small to contain in comfort a concourse of six thousand guests, of whom two thousand were said to have come uninvited.

The ball was thus, in the eyes of *The Times* correspondent, 'a singular and picturesque but not very "enjoyable" entertainment. Among the thousands were only two hundred women. There was a shortage of carriages, and many of the guests, like the players in *Hamlet*, came on their asses'—among them the Austrian Imperial Chancellor. Finding no conveyance, he started to walk, his feet sinking deeper into the sand at each step until he came upon a donkey, and proceeded to the palace astride it, 'making the best of his way to wait on his Kaiserlich Königlich Majesty'. It was a mob, 'pushing, squeezing, thick, suffocating. . . . From room to room men buffeted each other, in the vain hope of escaping from the press each did his best to create.' But outside in the fresh air, sweet with the scent of Ismailia's roses, 'the sand of the streets, soft to the foot, shone like a sheet of snow, as the light blazed on the particles of mica on the surface . . . and the camps were as if on fire'.

Amid the sand-dunes a huge supper-room had been contrived, with long tables laid for a thousand and an enclosure at the end for the royal and official visitors, transformed into a lantern-lit tropical garden, with plants brought from the horticultural Gardens of Gezireh in Cairo. The menu of twenty-four dishes included *Poisson à la Réunion des Deux Mers, Roast Beef à l'Anglaise*, and a salad of *Crevettes de Suez au Cresson*. The meal was not served until one o'clock in the morning, when 'savage famine could alone excuse the conduct of the ferocious multitude. Every seat at table was occupied before the supper hour had arrived, and the people acting on the principle followed by the cripples at the

[1] January, 1870.

Pool of Siloam, who saw that it was only those who pushed who got anything, certainly presented a scene from civilized Europe which would have astonished any onlooking Beduin.'

*   *   *

On November 19 the voyage continued, first to the Bitter Lakes and thence down the last stretch of the Canal to Suez. The way out of Lake Timsah, leading into the narrow deep cutting of the Canal, was marked by flags on posts. As the *Aigle,* leading the convoy as before, approached the Serapeum, with its sunken ledge of rock, there was a feeling of tension on board. But she passed over it easily, and as she did so the workers on the banks— men whose dredgers had been clearing the channel right up to the last moment and would resume their work as soon as the ships had passed—visibly expressed their relief and satisfaction. Soon the ships were sailing into the Bitter Lakes, this 'inland sea, with waves crisping in the breeze, under the bright sun, filling the whole Desert with the joyous noise of rushing water'. As *The Times* correspondent, with his fellow-passengers in the *Cambria*, sipped symbolically and circumspectly from a bumper of its bitter and saline water, he reflected: 'If the man who discovers a new island is honoured, he who creates a navigable lake in a lifeless waste, which shall henceforth be laid open to the eyes of wondering peoples as they pass to and fro, is still more entitled to honour.'

The *Aigle* lay for the night in the Bitter Lakes, with fifteen other vessels, saluting each other with rockets as their royal passengers exchanged visits after dark. Next morning, passing easily through the troublesome cutting of Shallufa, where men were still at work, the convoy saw 'the veritable Red Sea' ahead of it, and was soon sliding into this 'narrow neck of water which springs from the vast oceans that wash India and China and Western Africa'.

The *Aigle* rounded the end of the Canal embankment to an Egyptian salute of twenty-one guns, soon echoed by the Fort and by the naval vessels which crowded the roadstead. The Egyptian troops on the jetty presented arms to the Empress. The sailors lining the shore gave her the Egyptian equivalent of a round of cheers, with a repetitive staccato bark of 'God protect you! God protect you!' The bands on the ships struck up *'Partant pour la Syrie'*.

When the *Aigle* had anchored, the Khedive in his State barge was rowed out to her by twelve oarsmen, and climbed on board to greet the Empress. Meanwhile, in slow and solemn succession, the royal and other ships sailed into Suez, 'moving in single file, their hulls hidden by the bank down the Canal, as if they were coming by railway'. They had covered the whole hundred miles of it in an average sailing time of sixteen hours.

The British Admiral in Command sent a signal to the Admiralty: 'Empress, Psyche, Newport arrived. Canal is a great success.' Following it up with a despatch, he recorded: 'The arrival of about thirty-five ships in the Red Sea from Port Said drawing from ten to seventeen feet water has established the passage of the Canal, which is a work of vast magnitude, conceived and carried out by the energy and perseverance of M. Lesseps.' Mr. Elliot, the British Ambassador, confirmed the news to the Foreign Office. He agreed with the Admiral that much work still remained to be done, in terms of deepening and widening, before the Canal would be ready to take the larger vessels for which it was designed. Nevertheless, the opening had been a complete success, and 'this vast undertaking may now be regarded as triumphantly accomplished'.

The Foreign Office, on behalf of Her Majesty's Government, sent a telegram of congratulation to Lesseps on 'the establishment of a new channel of communication between East and West'. His 'indomitable perseverance' in surmounting obstacles, 'the necessary result both of physical circumstances and of a local state of society to which such undertakings were unknown', had been finally rewarded by 'a brilliant success'. It was a moment for Lesseps, in composing a gracious reply, handsomely to forget those other and more formidable political obstacles, which had all but brought the project of the Suez Canal to grief.

Having reached Suez, the Crowned Heads departed by train with the Khedive to Cairo, where festivities continued for a further week. The Empress remained at Suez for a day, resting and visiting, in the steps of the first Napoleon, the Wells of Moses on Sinai. She was delayed, as he had been, and returned too late to perform the unveiling ceremony at Port Tewfik of a bust of Thomas Waghorn—that earlier pioneer whose 'heroic perseverance', in Lesseps's words, 'had triumphed over the most incredible difficulties to get adopted by England the mail route to India across Egypt'. The Empress, however, found time to visit a house by the sea which Napoleon was said to have inhabited. It had been bought by a Moslem, a strong Bonapartist, who remarked in his honour: 'Buonaberdi could have burned all the mosques of Egypt. He did not do so. Let his name be blessed!'

On November 22 the *Aigle*, still with the Empress on board, made the return journey up the Canal to Port Said, spending the night on Lake Timsah. The next afternoon, seen off by Princes of the Khedivial house and by Lesseps, she left for France, where in the following fatal year of 1870 the cheerless destiny of dethronement and exile awaited her.

On the day after her departure Ferdinand de Lesseps, 'having united two oceans and converted Africa into an island'—as *The Times* put it—'terminated his labours appropriately by marrying a young and charming lady'. His bride was Louise-Hélène Autard de Bragard, daughter of an old friend in Mauritius, whom he had first met at one of the Empress's weekly receptions, and who had been his principal personal guest at the opening ceremonies. She was twenty years old and her bridegroom sixty-four. They were married privately by the Protonotary Apostolic at Ismailia, and she was to bear him twelve children (to add to the two surviving sons of his previous marriage) between now and his death at the age of eighty-nine—a few days after the twenty-fifth anniversary of the opening of the Suez Canal.

With his bride he paid a triumphal visit to England, where Queen Victoria invested him with the appropriate Order of the Grand Cross of the Star of India—the dominion which he had brought so much nearer to Britain. The Prince of Wales conferred on him the Gold Albert Medal of the Society of Arts. He was made an honorary freeman of the City of London, whose

Chamberlain, at a banquet in the Guildhall, declaimed amid cheers: 'You have succeeded in subjugating to your will the intractable sand, supplanting the patient, much-enduring "ship of the desert" by the vessels of all nations bearing the commerce of the world.' The honoured guest had, in short, 'rectified geography'.

Everywhere he was generously received. At a banquet in his honour at Stafford House, Mr. Gladstone, the Prime Minister, proposed his health, quoting the dramatist's words:

> Ye Gods, annihilate but time and space,
> And make two lovers happy.

Lesseps had achieved, if not 'annihilation' then 'abbreviation', in joining two seas, a feat rivalling that of the explorers who had crossed mountains and traversed oceans to discover the world's geographical surface. 'To connect distant places and bring the inhabitants of the globe into easier and more frequest communication was now the great want of science and the great work of the discoverer,' and Mr. de Lesseps had achieved this.

Lesseps in reply recalled how Gladstone, on his earliest visit to Britain, had said to him: 'Do not preoccupy yourself with the difficulties you may have to encounter in our country or in other countries. Continue your work with perseverance; and when you have succeeded the Suez Canal will be so useful to England that you are sure to meet in our country with the greatest support. . . . It is here that you will be crowned with glory.'

His prophetic words were now handsomely realized.

CHAPTER XVI

'You have it, Madam!'

The Khedive's honoured guests had no sooner dispersed than he received a sharp call to order from the Porte, with which his relations had progressively worsened. Clouds from Constantinople had indeed darkened for Ismail the inaugural ceremonies, causing him to seek earnest consultations with the British Ambassador and other diplomats present.

The storm had been gathering since the summer when, on his return from Europe, the Khedive was confronted by a virtual ultimatum from the Grand Vizier, designed to curb his insubordination in the military, diplomatic and above all financial spheres. The Porte was chiefly concerned at his contraction, without its permission, of foreign loans at high rates of interest, on an increasingly irresponsible scale. Such extravagance, it was feared at the Porte, was likely to prejudice the payment of the annual tribute from Egypt, on which its own slender revenues counted, and moreover compromise its chances of obtaining much-needed European loans for itself.

Above all it seemed to threaten a degree of insolvency in Egypt which could put her in the power of the foreigner, at the expense of Turkish influence, and to the detriment of the credit and prestige of the Ottoman Empire in general. To prevent such a crisis was likewise in the interests of the European powers, who were disturbed at Ismail's mounting taxation and consequent oppression of the people, his reckless speculations and losses, especially in cotton, and his expenditure on military excursions up the Nile with a view to Central African conquests.

The Khedive returned an unsatisfactory reply to these various strictures. At the end of November 1869, on the conclusion of the inaugural ceremonies, the Porte answered him back with an

Imperial firman. In this the Sultan insisted on his sovereign duty and right to ensure that the proceeds of taxes levied in his name should be devoted to the true interests of Egypt, and not to unfruitful expenditure. As to foreign loans, which committed the country's revenues on a long-term basis, these could only be sanctioned in cases of absolute necessity. The exact purposes for which they were needed must first be put forward in detail to the Imperial Government and their contraction must depend on its prior authorization. Ismail was instructed henceforward to conform with these terms, and a Commissioner was sent to Egypt to deliver the firman and ensure his compliance.

Angry and humiliated, the Khedive made a show of receiving the firman without accepting it, thus saving face, while the face of the Commissioner was saved by its public announcement. In a correct but non-committal letter of acknowledgement to the Grand Vizier, Ismail foreshadowed demands which he intended to place at the feet of His Gracious Majesty at some future date. Under the moderating influence of the Ambassadors, he made certain concessions. Then he intensified his policy of bribery and intrigue at the Porte which was to culminate, three years later, in a new firman—under a new Grand Vizier—followed by another, which granted him much of the freedom he sought in financial and foreign affairs. After a brief period of restriction, in which he fell back on the ready financial assistance of the local Levantine market, the Khedive was able to resume his march to ruin, with the aid of larger foreign loans than ever.

*   *   *

Meanwhile, in its hour of triumph, the Suez Canal Company itself had run, once more, into acute financial trouble. The Canal had proved, despite Palmerston, physically practicable. But was it also, despite him, to prove financially profitable? This had now to be put to the test. The shareholders, over-encouraged by Lesseps throughout these years, were now expecting an immediate return on their money, while Lesseps was counting on an early revenue of 10 million francs to cover not only the dividends of ordinary shareholders and subsequent bondholders, but an anticipated capital deficit due largely to arrears of interest on the coupons of the shares.

But Europe, conservative in the matter of its trade-routes, proved slow to take advantage of the new Canal. Lesseps might write to his son that he 'never tired' of the spectacle of fleets crossing the desert, that he would gallop on horseback to a point of vantage to watch a big steamship pass, that he was stirred to the heart when the passengers and crew gathered on deck to cheer him. But in fact, during the first year of its operation, only some five hundred ships passed through the Canal, bringing in a revenue of a mere 4 million francs. Thus the Company could not meet its obligations.

A clamour arose among the shareholders, alarmed and indignant at the delay in their interest payments. The shares dropped sharply, and the prospect of liquidation was freely discussed in the French Press. To meet the deficit and tide the Company over these first lean years, Lesseps announced to the Annual General Meeting in 1871 a projected loan of 20 million francs, for which the Khedive would provide cover by permitting the levy of a surtax of one franc per ton above the normal rate of Canal tolls.

But Lesseps had lost much of his sway over the shareholders, who interrupted him constantly and contested his financial report. Public confidence had fallen so low that there was at first little response to the loan. Nor was official support now forthcoming. But it was kept open for a further six months, and early in 1872 12 million francs were subscribed. This gave the Company a breathing space, and enabled it to pay the arrears of interest over the next two years. The coupons were capitalized, in the form of bonds, paying 5 per cent interest and redeemable in forty years.

To increase revenue Lesseps now tried to obtain an increase in the tolls, to which shipping was liable for the use of the Canal, and which had been laid down by the Company's rules on the basis of ten francs per ton. But the relevant clause in the Concession was far from explicit. The fixing of the tariff thus raised complex issues, involving the definition of tonnage and the methods of its calculation adopted by different countries.

At first the tolls were calculated in terms of the net tonnage of ships, allowing for the deduction of space occupied by engines and fuel. But in 1872 the Company announced that in future it would be calculated on a basis which amounted in effect to gross tonnage. This was based on the Company's own assessment, with the advice of a Commission of engineers and shippers, of the space to be deducted, which conflicted with the prevailing system in Britain and elsewhere. It involved an increase of some 30 per cent above the previous rate.

This led to protests from shippers, arguments with Governments, litigation in the French courts, disputes on jurisdiction, as between French and Egyptian Courts, and finally, in January 1873, the convocation by the Porte of an International Conference of the maritime powers at Constantinople. Lesseps fought stubbornly in defence of what he took to be a fundamental principle of his Company's legal and moral right—to say nothing of its urgent financial need—to fix its own scale of tariffs. He was back at what Elliot, the British Ambassador, described as 'his favourite game of menace and intimidation . . . putting his case before the public with great ingenuity and perseverance through the local press, which he has secured to his interest'; acting as though he were himself a 'supplemental Ambassador'; and accusing Her Majesty's Government of opening a 'new war between England and France'.

To Sir Daniel Lange, his representative in London, he wrote: 'We can only reply to those who are not satisfied with our terms for the passage through the Suez Canal that they are at liberty to avail themselves of the Egyptian Railway or if they prefer doing so go round the Cape of Good Hope as before.'

The Conference none the less decided against him, thanks largely to the influence of Britain, backed by Austria, Germany and Russia, and opposed only by France. It settled for the British system of tolls, which corresponded more exactly to international usage. But it granted a concession to the Company in the form of the right to levy a temporary surtax of four francs per ton on the net tonnage. Lesseps, protesting, proposed a continuation of this surtax until the shareholders had received the interest due on their coupons, until improvements in the Canal had been completed

Panorama of the Canal from sea to sea; an artist's impression

The Khedive rows out to meet the Empress on arrival at Suez

Illuminations at Port Said

and until the net revenue on its capital had reached 8 per cent. The Porte's reply to this was intransigent. The Company was warned that if it did not comply with the decision, there would be a reversion to the original toll of ten francs per ton, without any surtax at all. This ultimatum provoked an explosion from Lesseps. To Nubar he affected incredulity that he should, after all these twenty years, be treated as though he had committed a crime instead of rendering a service to humanity. To the Khedive he declared that he would hold the Porte responsible for all losses to shareholders, amounting to 700,000 francs a month, which promised to result from this regulation. He even went so far as to threaten to abandon the Canal altogether, to extinguish the lights, remove signals, refuse pilots, cut telegraphic communications. Meanwhile, on the Canal itself, the Company's officials had adopted an obstructive and high-handed attitude.

The Khedive's reply to this was to send a large military force, following orders from the Porte, now certain of British approval, to the Canal, and two warships to Port Said, with instructions to seize the Canal if the Company's opposition persisted. France denied Lesseps further support, and he accepted defeat. He had been forced to abandon his unilateral pretensions and conform to an international system. But for the present he lost little in practice, since the rates which he now had to accept amounted, with the surtax, to only a little less than those which he had tried to impose.

*   *   *

This crisis had brought into the open a new and significant political element—the increasing interest and influence of Britain, at the expense of France, in the affairs of the Suez Canal. The disastrous war of 1870 had brought down the Second Empire, driving the Emperor and the Empress Eugénie into exile, and had initiated the Third Republic. The change of régimes brought a change in official attitudes towards the Canal. Under the Empire the French Government, concerned not to antagonize Britain, generally withheld overt support for the Canal Company, but was willing and able to support it, when needed, in practice. Now this trend was reversed. Under the Republic the Government, concerned to protect French investors, gave the Company overt support but was often unable, and sometimes unwilling, to implement it in practice.

Such, following France's defeat, was the measure of her loss of weight and international authority in Europe and in the Middle East, from which Britain now profited. Britain had consistently opposed the Canal as a French imperial threat, before it was made. But now that it was made she determined to convert it into a British imperial asset—a task made easier by the fact that France was no longer an Empire. Before the Canal's completion, Lesseps had declared to his shareholders: 'France will have subscribed the greater portion of your capital but England will pay you the largest proportion of your dividends.' And as revenue now slowly mounted, year by year, with the increase in the number of ships making use of the Canal, this became even more evident. For two-thirds of their tonnage was British. As the Board of Trade was soon to report: 'The trade between Europe and the East flows more and more through the Canal, and the British Flag covers an ever increasing proportion of this trade.'

Logically this should lead to an increasing measure of British control over the Suez Canal. But Britain's Liberal Government was moving with caution. At the end of 1870 Major-General Stanton (as he now was), the British Consul-General, reported to London an important proposal from the Khedive. Alluding to the French Company's financial difficulties, Ismail observed to him that of all countries Britain was the most concerned to keep the Canal open. This being so, he told the British agent that 'he would gladly see the Canal the property of an English Company, and that in the event of such a Company being formed he would do everything in his power to facilitate the transfer into their hands'.

Expressing his own opinion to Lord Granville, the Foreign Secretary, Stanton commented: 'A great opportunity may shortly occur of consolidating British influence in this country and at the same time securing our communications with India by obtaining possession of the Suez Canal, a possession which would in my opinion be attended by most important political advantages to Her Majesty's Government.' To this suggestion Granville did not reply. But he tried it out on the interested Government departments, prompting comment that was favourable from the India Office but at first critical from the Board of Trade.

A few months later, in April 1871, the same proposal, in a more positive form, reached Granville from another quarter—within Lesseps's own camp. Its author was Sir Daniel Lange, the Canal Company's representative in Britain. The Canal, Lange reported, had just been finally completed, to a uniform depth, and could now take the largest ships. But it remained to be seen how the shareholders, at their annual meeting later in the month, would react 'when their patriotism is again invoked, for the purpose of submitting to further pecuniary sacrifices, rather than assent to the alternative of British assistance and control in the future management of the Suez Canal'. He suggested that 'a well-defined financial proposal emanating from England' might be 'opportune and even acceptable to the shareholders then, provided it be done with discretion and caution'. He preferred first to ascertain Lesseps's own views. For, after all, 'we have worked together to unite the two seas. He "for the glory of France". I for the interest of England.' None the less it was surely 'desirable that the Suez Canal should become, by just means, subject to the control of this country'.

Later in the month Lesseps was in London, trying to raise money for his 20 million franc loan. Lange, as he wrote to Granville, suggested such a course to him, 'as one likely to be permanent in its results and not exposed to the necessity of further makeshifts such as temporary loans, which would merely serve to stave off an evil for the time being'. He said openly to Lesseps, 'Place before the shareholders the alternative of no dividends or British assistance and entire management'—with himself still as President in Paris. But Lesseps 'recoiled with aversion from this proposal and declared he never would be a party to the transfer of the management of the Canal into other than French hands'. All he would accept would be the introduction on to the French Board of a few English directors.

Lange, though hardly surprised at this attitude, believed it to be based on a dangerous over-confidence. He feared, as he wrote to Granville, that 'the present fragile tenure of the Canal might break down for want of funds, and leave the management in the hands of persons to whom it was never intended to confide it'. It was not until three years later that Lesseps learnt of this correspondence with the Foreign Office by his trusted British henchman. Enraged at such disloyalty, he secured Lange's immediate dismissal.

Meanwhile Lesseps, in so far as he was prepared to consider a sale, preferred the idea of purchase, on specified conditions, by the Governments of several maritime powers, thus giving the Canal an international character. A plan of this nature was launched in Italy, with Austrian support, and gained favour with the French Government. Granville however found it premature to give an opinion. The Khedive was unimpressed by the proposal, which seemed to him to raise considerable difficulties. But he passed it over to the Porte. The Porte declared that Lesseps had no right to sell the Canal. And the idea was gradually dropped.

In fact, from the end of 1873 onwards, the position of the Company began to improve. Traffic increased appreciably. The settlement of the tonnage dispute, with its authorization of the surtax, brought in adequate revenue. In 1874 the Company was able to launch another bond issue, this time for 35 million francs, and it was fully subscribed. Thus all the arrears of interest were paid, and the market value of the ordinary shares reached a premium. Ugly rumours of liquidation receded. The Suez Canal seemed to be weathering its first lean years.

*   *   *

The Khedive, on the other hand, was now heading towards a crucial stage on his profligate road to financial disaster. In 1872 he had contracted the most expensive of all his foreign loans, for 800 million francs (£32 million). By 1875 his public debt had reached almost £100 million, of which £68 million was owed abroad. The annual interest on his foreign loans was some £5 million, more than the total income of Egypt in the year of his accession. Thus most of the money now borrowed was swallowed up by the interest on previous loans, and there remained little to show for it. 'For all practical purposes', as Lord Cromer wrote later, 'it may be said that the whole of the borrowed money, except £16,000,000 spent on the Canal, was squandered.'

Now his Ordinary shares in the Canal Company, amounting to roughly a quarter of this total sum, were virtually his only liquid asset. In November 1875 he found himself in urgent need of three or four million pounds sterling, to meet obligations due at the beginning of December and in the early months of 1876, and thus to stave off a probable suspension of payments. It became clear that he would be obliged to sell or mortgage these Canal shares, 177,642 in number.

Two rival groups of French bankers, with extensive claims on the Egyptian debt, were soon contending in Paris, one for the purchase of the shares, the other for their retention as security for a mortgage. First in the field were the brothers Dervieu, Edouard in Paris and André in Alexandria, who had been closely involved in viceregal finances since Ismail's accession, and who foresaw this crisis early in November. On Friday, November 11, they approached the Khedive, who agreed to sell them his shares for 92 million francs (£3,680,000) and to pay them 8 per cent interest—later raised to 11 per cent—secured on the Customs of Port Said and on the coupons of the shares, which he had signed away to the Company for a further nineteen years. Requesting time to raise the funds, they were granted an option until November 16. They had close links with the *Société Générale,* who undertook to participate in the purchase but not to promote it.

But meanwhile, as Edouard Dervieu soon discovered, another syndicate was in the field, with the backing of the *Crédit Foncier,* which was represented in Egypt by the Anglo-Egyptian Bank. Its objective was not the purchase of the Canal shares but the establishment of solvency in Egypt through the conversion of its floating debt, composed mostly of short-term obligations, into a consolidated long-term loan. For this the subscribers would demand guarantees, hence the Khedive would be required to retain his shares. The two approaches were thus incompatible, and Dervieu had to seek elsewhere, meanwhile obtaining an extension of his option until the following Friday, November 19.

Stanton, the British Consul-General in Cairo, knew nothing of these operations until Monday, November 15, when he received a cable from the Foreign Office, enquiring urgently whether the news of such an offer, by 'a combination of French capitalists' was true. London had first got news of it over the weekend, when Henry Oppenheim, of the Paris banking house, for long an associate but later an adversary of Dervieu in Egypt, revealed it at dinner on Sunday, November 14, to Frederick Greenwood, the proprietor and editor of the *Pall Mall Gazette*. Oppenheim, a major creditor of the Khedive, was associated with the *Crédit Foncier*, hence opposed to the Dervieu plan and concerned to influence the British Government against it.

On Monday morning Greenwood passed on the news in person to the Foreign Secretary, Lord Derby, with a strong recommendation that the British Government should purchase the shares for itself. A Tory administration was now in power, with Disraeli as Prime Minister in succession to Gladstone. Disraeli may indeed have heard the story already from Baron Lionel de Rothschild, who knew all that passed in the financial world abroad, and with whom Disraeli was in the habit of dining on Sunday evenings. For at his house, as he put it, 'there is ever something to learn'.

The possibility of acquiring an interest in the Canal for Britain had occurred to Disraeli soon after the Tory Government came to power in 1874, and he had sent Baron Rothschild's son on a confidential mission to Paris to enquire into the prospects. But nothing came of his enquiry, since Lesseps refused any sale— indicating his refusal by naming a price, £40 million, so prohibitive that it 'startled even a Rothschild'. When, on June 5, 1874, Lord Derby was questioned in the House of Lords as to whether, in view of the Company's position, it would not be advisable for Britain to purchase the Canal, he replied that 'if a proposition for the transfer of the Canal to an International Commission were to come before us, framed in such a manner that all Governments would participate in the advantages of the Canal on equal terms, I do not say that might not be a fair proposal to entertain. But it has not been made, and I have no reason to think it will be made.'

Now all of a sudden, from another quarter and in different terms, the prospect of such a purchase materialized. Here was the chance of a political stroke which appealed instantly to Disraeli's imagination and to his sense of Britain's imperial interests. But there was no time to be lost. From Stanton a reply came on Tuesday, giving news of two offers and stating that he had secured from the Khedive a suspension of negotiations until Thursday. On Wednesday the Cabinet met. Disraeli found that his Ministers did not all at first share his enthusiasm. Derby maintained a cautious reserve. Sir Stafford Northcote, the Chancellor of the Exchequer, had scruples on the grounds of political morality, writing later to Disraeli: 'We opposed it [the Canal] in its origin; we refused to help Lesseps in his difficulties; we have used it when it has succeeded; we have fought the battle of our shipowners very stiffly; and now we avail ourselves of our influence with Egypt to get a quiet slice of what promises to be a good thing. I don't like it.'

But in view of the Prime Minister's determination, the Cabinet unanimously backed him, deciding in principle that Britain should acquire the shares. A telegram was despatched immediately to Stanton: 'It is of great importance that the interest of the Viceroy of Egypt in the Suez Canal should not fall into the hands of a foreign company. Press for suspension of negotiations, and intimate that Her Majesty's Government are disposed to purchase if terms can be arranged.' That evening Stanton saw the Khedive, who assured him, in response to this communication, that he had no present intention of selling his shares, but that if he changed his mind he would give Her Majesty's Government the option of purchase. Meanwhile he was considering a proposal for a loan, on mortgage, with the shares as security.

Next day Disraeli communicated the news to Queen Victoria: "'Tis an affair of millions; about four at least; but would give the possessor an immense, not to say preponderating, influence in the management of the Canal. It is vital to your Majesty's authority and power at this critical moment, that the Canal should belong to England, and I was so decided and absolute with Lord Derby on this head, that he ultimately adopted my views and brought the matter before the Cabinet yesterday. The Cabinet was unanimous in their decision, that the interest of the Khedive should, if possible, be obtained, and we telegraphed accordingly.' The Queen telegraphed her approval from Balmoral, but feared the matter might be difficult to arrange.

The Queen's representative in Egypt, Major-General Stanton, 'as honest a soldier as ever represented England abroad' (as one of his contemporaries described him),[1] was sometimes out of his depth in the shoals of diplomatic and financial intrigue which beset foreigners at the Court of the Khedive—more than ever at this time, when desperate expedients were needed to stave off the spectre of imminent ruin. No sooner was it known that the British Government was in the market for the Khedive's shares than the two Pashas, Nubar and Sherif, born bargainers both—to say nothing of Ismail himself—made it their object to bid up the price. In Stanton, not well-informed as to what was afoot in the world of Egyptian finance, they found a customer ready enough to fall for their stratagems. They conveyed to him the impression that both groups of French bankers were actively in the market, competing with one another and thus now with the British Government for a deal with the Khedive—Dervieu and the *Société Générale* for their purchase for 90 million francs; the Anglo-Egyptian Bank and the *Crédit Foncier* for their retention as cover for an advance of 80 million francs, in terms of the consolidation of the debt.

In all probability the Anglo-Egyptian Bank, though interested with the *Crédit Foncier* in the long-term funding of the debt, made no such specific mortgage offer, being mainly concerned at this stage to prevent any purchase deal by the *Société Générale* or others. But in the swift exchange of telegrams which passed between London and Cairo throughout this crucial week— 'scarcely breathing time', as Disraeli put it, 'but the thing must be done'—Stanton acted in the belief that the offer had been

[1] C. F. Moberly Bell: *Khedives and Pashas.*

made, reporting indeed that the Khedive was on the point of accepting it. His report prompted an urgent query from Derby as to whether such an acceptance would preclude redemption of the Khedivial shares, followed by enquiries for details of the offer and of the identity of the Cairo bankers concerned with it.

Then, on Saturday, November 21, it suited Nubar to inform Stanton that the offer had been rejected, on account of its excessive demands, but that a new mortgage offer, from the *Société Générale*, was now being considered. This was the work of André Dervieu, whose brother Edouard had failed, thanks to the opposition of the *Crédit Foncier*, to raise in Paris the funds required for his original offer of purchase. Instead, on November 18, he had proposed to the Khedive not a purchase but a short-term loan amounting to 80 million francs (stepped up by Nubar for Stanton's benefit to 85 millions) with interest at 18 per cent, secured not only on the Canal shares but on the Khedive's statutory 15 per cent interest in the Company's profits. In the event of non-repayment within three months, both these assets would revert to the French syndicate. Meanwhile the Khedive would pay 10 per cent interest for nineteen years on the alienated coupons of the shares, secured, as proposed before, on the Customs of Port Said.

To raise the funds for this altogether favourable deal, Edouard Dervieu now had recourse to the French Government. He found an active intermediary in Lesseps himself. The proposal, as Lesseps pointed out to the Duc Decazes, the Foreign Minister, amounted in practice to a purchase for French interests of the Khedive's shares and of his profits in the Company, since he would clearly be unable to repay the loan within the stipulated three months. But the Minister of Finance opposed the scheme, on account of the French Government's close association with the *Crédit Foncier*; hence his support for the rival conversion plan. The Duc Decazes too 'resisted the persuasive and really seductive addresses of M. de Lesseps'—but for political reasons.

With Germany threatening to renew the war against France, and the British Government restraining her, he was reluctant at this moment to take any step which might prejudice French relations with Britain. Thus he instructed Gavard, his Chargé d'Affaires in London, to test the British official reaction to a possible French purchase of the Khedive's shares. Lord Derby's response, as reported by Gavard to Paris on November 20, was emphatic:

> I do not conceal from you that I should see serious incon-
> venience in such a course. You know what my opinion is
> regarding the French Company. It has run all the risks of the
> enterprise; all honour is due to it. . . . But since we use the
> Canal more than all the other nations put together . . . I should
> be very glad to see the time come when it would be possible to
> buy out the Shareholders and replace the Company by a
> Syndicate or Administration in which all the Maritime Powers
> would be represented. In any case we will do our utmost not to
> let an undertaking on which so many of our interests depend
> be monopolized by foreigners. The guarantee resulting from the
> control of the Porte is now no longer sufficient. If we lost that
> offered by the participation of the Khedive we should be abso-
> lutely at the mercy of M. de Lesseps. The French Shareholders
> already possess 110 millions out of the 200 million [francs]
> which the capital of the Shares represents. It is enough.

Decazes' reaction to this was conclusive. He refused French official support. Dervieu and his syndicate renounced their project. The way lay open for Britain. In Cairo a last screw was turned by the Khedive. Sherif, on his behalf, pretended men-daciously to Stanton that he had received an offer of purchase of his shares from Lesseps, for 100 million francs, supposedly on account of the French Government. The Khedive, he said, pre-ferred to sell them to the British Government, and would settle for the same amount. In reply to Stanton, on November 24, 1875, a week after the Cabinet first discussed the purchase, Lord Derby telegraphed: 'The Viceroy's offer is accepted. Her Majesty's Government agree to purchase the 177,642 shares of the Viceroy for 4 million pounds sterling and to recommend Parliament to sanction the contract.'

In thus selling his shares to the British Government for the equivalent of 100 million francs, with an interest of 5 per cent until the coupons fell in, the Khedive had secured a good bargain

—better than he would have secured from the French syndicate, with its successive offers of 92 million and 85 million francs, involving him in higher rates of interest. The British purchase moreover did not include, as the second French bid would have done, the Khedive's permanent 15 per cent share in the Canal's profits. A week later he offered to sell this too, for another £2 million. But the British Government, largely in his own political interests and those of the Porte, rejected the offer. For its own part, the Government could be content with an excellent long-term investment. Within six years its shareholding rose in value from £4 million to £5,750,000. By the turn of the century it rose at the rate of £2 million a year until just before the outbreak of the First World War it was worth ten times as much as the original purchase price.

Meanwhile Disraeli broke the news of his masterly stroke to the Queen: 'It is just settled: you have it, Madam. The French Government has been out-generalled. They tried too much, offering loans at an usurious rate, and with conditions, which would have virtually given them the government of Egypt. . . . Four millions sterling! and almost immediately. . . . The entire interest of the Khedive is now Yours, Madam.' The Queen (who was to receive him at Windsor next day) replied that this was indeed 'a great and important event' which she felt sure would be most popular in the country. 'The great sum', she added, 'is the only disadvantage.'

Disraeli, as accomplished a showman as Lesseps and as much given to hyperbole, wrote also to Lady Bradford: 'We have had all the gamblers, capitalists, financiers of the world, organized and platooned in bands of plunderers, arrayed against us, and secret emissaries in every corner, and have battled them all, and have never been suspected. The day before yesterday, Lesseps, whose company has the remaining shares, backed by the French Government, whose agent he was, made a great offer. Had it succeeded, the whole of the Suez Canal wd. have belonged to France, and they might have shut it up! . . . the Faery is in ecstasies.'

After his audience he wrote again, confirming the Faery's excitement and adding: 'What she liked most was, it was a blow at Bismarck.' In fact Prussia was to see the deal rather as a blow at France. To the Queen's daughter, the Crown Princess, the purchase was '*the right thing* done at the *right moment*', while the Crown Prince (the later Kaiser Wilhelm II) wrote of his delight with the exclamation: 'How jolly! !'

In Britain the news of Disraeli's audacious *coup* aroused keen patriotic enthusiasm. According to *The Times*, this was one of those occasions 'on which the national acquiescence in an act of the Executive is given spontaneously and instantaneously . . . The news of the present purchase was received, as it were, with a start of pleasure . . . It appealed to a feeling far stronger at the present time than Constitutional prudishness or commercial caution.'[1] From Disraeli (in the words of the *Quarterly Review*)[2] 'the country expected some departure from commonplace. . . . The lion was shaking the dewdrops from his mane. His drowsiness had at last come to an end. If nothing more, he had given a growl of awakening, an *adsum*, an *acte de présence.*'

Misgivings were voiced only at the dangerous principle of a Government thus engaging in trade. *The Times*, quoting them, asked: 'If Britannia sits as a directress in one board-room, or interpellates in one meeting of Continental shareholders, why not in another?' There were those who feared that this 'unprecedented application of State funds' might lead Governments into such enterprises as 'railways, steam-boat lines, coal mines' and so produce 'a change in our political habits'. But *The Times* dismissed such fears.

*   *   *

Parliament was not at present in session. Had it been so the deal could not have gone through with such secrecy and speed, as Disraeli well knew. 'We cd. not call them together for the matter', he wrote to Lady Bradford, 'for that would have blown everything to the skies, or to Hades.' Thus Disraeli, as soon as his Cabinet had decided on purchase in principle, had taken steps to raise an immediate loan, covering the payments which would later have to be sanctioned by Parliament. For this he turned to his friends the Rothschilds.

[1] December 3, 1875.   [2] October, 1876.

His Private Secretary, Montagu Lowry-Corry (afterwards Lord Rowton) liked to tell a picturesque story of how this was done. By arrangement, he took up his post just outside the Cabinet Room. In the course of the meeting, the Prime Minister put his head out of the door and said, 'Yes'. At this pre-arranged signal, Corry went straight to the office of Baron de Rothschild and told him that the Prime Minister wanted £4 million.

'When?' asked Rothschild.

'Tomorrow.'

Rothschild, according to Corry, then picked up a muscatel grape, ate it, spat out the skin, and asked:

'What is your security?'

'The British Government.'

'You shall have it.'

Within a few days the details were arranged. Rothschilds agreed to provide the money in three instalments—£2 million on December 1, £1 million on December 16 and £1 million on January 5. They charged commission at the rate of 2½ per cent. The Treasury was disconcerted at so high a charge, which amounted to £100,000, or an average interest rate of thirteen per cent per year. The Bank of England was dismissed as a possible alternative source for the loan, because its legal powers for such a purpose were doubtful and it could not have acted in secrecy.

Thus by an ironic twist of fortune it was the Rothschilds, spurned by Lesseps in Paris twenty years earlier with their offer to handle the initial issue at 5 per cent, who now backed the transfer to Britain of a major part of it, at the expense of France. Frenchmen were bruised and humiliated at this unforeseen British *coup*. The Duc Decazes was attacked in the Republican press as giving 'proofs of a blindness unequalled save in the worst days of Imperial diplomacy'. But on the whole French comment reflected, in the phrase of the Paris correspondent of *The Times*, 'a kind of melancholy admiration'.[1] At least, he added, it was inspired by 'the conviction that English policy is one of broad daylight, which distinctly avows what it aims at'. A leading Paris financial

[1] November 29, 1875.

newspaper found more cause for congratulation than fear in the fact that 'England had thus shaken off her political lethargy'.

Lesseps rose above his discomfiture to report to his shareholders: 'Today the English nation accepts in the Canal Company the part which had been loyally reserved for it at the outset; and if this act, once accomplished, can have a consequence, that consequence can in my eyes be, on the part of the British Government, no other than the renunciation of an attitude which for a long time has been hostile to the shareholders who founded the Maritime Canal.' He considered as fortunate 'this powerful unity of interest which will be established between French and English capital for the purely industrial and necessary peaceful working of the universal Maritime Canal'.

He could after all claim that at least half the shares now acquired had been initially held for other nations—if not indeed solely for Britain—and had only been unloaded on the Viceroy when no bid was made for them. For Britain their acquisition ensured an influence in the affairs of the Canal which, despite international safeguards, might have been denied to her if France had acquired them. As owner of all the shares, France might well have set up for its management a rigid and unwieldy bureaucracy on the French model, to the possible detriment of commercial efficiency and of political impartiality.

Still Britain had no official voice in the Company's management. For the Khedive, in signing away the coupons of his shares, had signed away also the ten votes that went with them. Here Lesseps immediately acted in a co-operative spirit. He sought and reached an agreement, on behalf of the Company, with a representative of the British Government, Colonel John Stokes of the Royal Engineers, who had represented it earlier in the tonnage negotiations at Constantinople. This restored the voting rights to the shares, giving the Government ten votes. It also provided for the appointment to the Board of the Company of three British directors, one-eighth of the total. This number was increased eight years later by the addition of seven unofficial British directors, and provision for a British Vice-President. Meanwhile, in a second agreement with Stokes, the Company undertook, during the next thirty years, to spend a minimum of £40,000 a year on capital improvements to the Canal, and within eight years, by a system of progressive annual reductions, to eliminate the present surtax on tonnage and revert to the basic tariff alone.

*　　*　　*

This agreement was well-timed to forestall criticisms in Parliament, which met early in February 1876 and at once debated the purchase. Disraeli, as he expressed it in his address of reply to the Speech from the Throne, had said to the Rothschilds: 'Will you purchase these shares on our engagement that we will ask the House of Commons to take them off your hands?' And now the House, happy to accept a *fait accompli* which enjoyed wide support throughout the country and to which even certain prominent Liberals gave tacit approval, proceeded to do so by a grant in Supply. The Opposition had none the less to protest, taking, as *The Times* saw it, 'a narrow and minute view of the subject, while the majority of their countrymen, though they see it more vaguely, yet send their glances instinctively over a wider field, and thus, as we think, form a truer judgment'.[1] Mr. Lowe, the former Chancellor of the Exchequer, 'moved from one detail to another as if the subject had no breadth or dignity whatever,' making a speech 'like one of those ill-built old cities, all lanes and alleys, in which the stranger always thinks he must be close to a main thoroughfare, but never finds one'.

Mr. Gladstone, when the debate was resumed on February 21, grumbled at inordinate length, enquiring persistently into the minutiae of Britain's legal rights and remedies, and thus the extent of her influence under the terms of the purchase; complaining that everything was being done in too much of a hurry; shocked that a British Government should thus deal with a private firm—and at such a high rate of commission; foreseeing endless future claims by the Canal on the Exchequer and thus on the taxpayer. Altogether, the Opposition pretended, we had made an expensive and deplorable bargain.

[1] February 15, 1876.

Northcote, the current Chancellor, contested the various charges: 'When we are asked what we have got for our money, I reply that we have got value for our money, and in the second place we have obtained influence in the administration of the Canal.' As to the nature of that influence, he quoted Lord Derby: 'If anybody doubts that the possessor of two-fifths of the shares in this Canal obtains no influence, it is as difficult to argue with him as with a man who says that two and two do not make four.'

Winding up the debate, Disraeli referred to the 'great complications' which Gladstone foreshadowed: 'We are here to guard the country against complications, and to guide it in the event of complications; and the argument that we are to do nothing—never dare to move, never try to increase our strength and improve our position, because we are afraid of complications is certainly a new view of English policy, and one which I believe the House of Commons will never sanction.' Concluding, the Prime Minister stressed that he recommended the purchase not as a commercial but as a political speculation, 'and one which I believe is calculated to strengthen the Empire. That is the spirit in which it has been accepted by the country, which accepted it though the two right hon. critics may not. . . . The people of England want the Empire to be maintained, to be strengthened; they will not be alarmed even if it be increased. Because they think we are obtaining a great hold and interest in this important portion of Africa—because they think it secures to us a highway to our Indian Empire and our other dependencies, the people of England have from the first recognized the propriety and the wisdom of the step which we shall sanction tonight.' It was sanctioned unanimously, without a Division, and thus, as *The Times* put it next morning, 'the nation enters peaceably on its new possession'.

*   *   *

The new possession, packed in seven zinc boxes, had been deposited in the British Consulate-General in Cairo as soon as the agreement with the Khedive was signed. Stanton signed a receipt for the boxes, which were corded and sealed with the seals of the Finance Minister, Her Majesty's Consul-General, and the Consular court, pending a final check on the number of shares they contained. There proved to be only 176,602, not 177,642 as was supposed, and Nubar Pasha now recalled that a few shares had been sold in Paris some twelve years before. Thus the agreement was amended and a sum deducted from the price, bringing it down to £3,976,582.

Baron de Rothschild requested that a Notary Public should make a list of the shares with their numbers and other particulars, with two certified copies for himself and Her Majesty's Government. But there were no Notaries Public in Egypt. Thus the list had to be made by the British Vice-Consul, with a delegate from the Egyptian Ministry of Finance. This was an assignment calling for immense time and labour, to which the Consulate's resources, unless it were to lay aside all other work, were unequal. Hence Stanton applied to London for authorization to engage and pay extra clerical staff. Even so, to save time, it was found advisable to register the shares not singly but in groups of fifty and a hundred.

When the task was completed the problem arose of transporting them safely to London. Should the Vice-Consul, as a special messenger, convey them home? Should one of Her Majesty's ships come to collect them? Now that the Canal was in use few mail steamers called at Alexandria, and Stanton was becoming concerned, not only at the responsibility of his charge, but at the amount of space which the seven cases took up in his office.

Finally the Admiralty, at the request of the Foreign Office, instructed the *Malabar,* a troopship passing home through the Red Sea, to make a special call at Alexandria and there 'receive certain cases'. On the day appointed, Stanton travelled down to Alexandria, in a special train placed at his disposal by the Khedive, and the precious cargo was stored on board. Arriving at Portsmouth on the last day of the year 1875, the seven boxes were swiftly conveyed, under guard, to the Bank of England.

And now, with Parliament's blessing, Egypt's capital interest in the Suez Canal was indisputably held by the Queen.

CHAPTER XVII

Britain moves in

Disraeli's purchase of the Canal shares was generally assumed, both as home and abroad, to be a prelude to some form of British control over Egypt. Liberals feared the worst, and voiced their fears in Parliament and elsewhere. Lord Derby, to Queen Victoria's irritation, poured 'cold water' on any such idea, insisting that the Government had acted simply to prevent this great highway of British shipping from being 'exclusively in the hands of the foreign shareholders of a foreign company'. Disraeli, perorating to the House of Commons, had not discouraged the people of England from playing with the idea. But in essence the aim of his Middle Eastern policy was still to forestall a French rather than to further a British occupation of Egypt, and to uphold the integrity of the Ottoman Empire.

The Suez Canal, none the less, was to call for the revision of such traditional policies. Lesseps, in changing the world's geography, had changed also not merely its commercial but its strategic and political alignments. Britain was switching fleets from the oceans, which she was accustomed to rule, to a narrow channel between lands over which she had no authority. Here she needed to ensure a security which the Sick Man of Europe could no longer provide. Inevitably Egypt would come to replace Turkey as a focal point in Britain's Middle Eastern policy. Inexorably events in Egypt—financial, political and finally military— marched from now onwards in the direction of British control, which in the end was, ironically, to be imposed by a Liberal Government.

The Khedive, in his imminent insolvency, was now driven to ask for a British expert to investigate and advise on his country's financial position. The British Government sent Mr. Stephen

Cave, M.P., who, in March 1876, recommended the consolidation of the Egyptian debt and the appointment of a Commission of financial control. Ismail thus established the *Caisse de la Dette*, with commissioners from four countries, and consolidated all his loans into a general debt of £91 million. This system was later supplemented by one of Dual Control, by Britain and France.

As the situation failed to improve, they obliged the Khedive, in March 1878, to accept the status of a constitutional ruler, with a Chamber of Notables and a Cabinet including a British and a French Minister. Persisting none the less in his former extravagant habits, seeking to exploit the Dual Control as a means rather to evade than to discharge his responsibilities, Ismail finally dismissed his Ministers and made a bid to resume despotic rule. But Britain and France, for once in unison, appealed to the Porte, and in June 1879 the Sultan deposed him. Thus Ismail disappeared into exile.

His son Tewfik, who now became Khedive, was soon confronted with a serious internal upheaval. When, later in the year, he reduced the native force in his Army and placed 2,500 officers on half-pay, a riot broke out in Cairo, which was to become the signal for a major Nationalist revolt. Its leader was an Army Colonel named Arabi, a fellah by origin; its slogan was 'Egypt for the Egyptians!'; and it counted among its supporters not only Army officers but educated civilians, resentful at the privileges accorded, over the past fifty years, not only to Europeans but to Turks and Circassians. This movement came to a head with a *coup d'état* in September 1881. Arabi, with 5,000 troops, surrounded the Khedive's Palace and compelled him to dismiss his Minister of War, whose place he took a few months later. Within a few months, after a brief reassertion by Tewfik of his legitimate powers, Arabi was Prime Minister. Thus Egypt was to be ruled for a spell by a régime which amounted to a military dictatorship.

The powers could not allow this to last. The European bondholders feared that Egyptian debts would be repudiated. The British Government feared for the security of the Suez Canal. France and Britain delivered a joint note to the Khedive, urging him to 'maintain and assert his proper authority'. This presumptuous intervention was deeply resented, not only by the Khedive but by the Sultan, who saw in it an encroachment on his own sovereign authority. Their reaction, in the eyes of Lord Cromer (now, as Major Evelyn Baring, a Commissioner on the Dual Control), made foreign intervention 'an unavoidable necessity'.

Anxious to avoid it, Britain and France despatched naval squadrons to Alexandria, to protect foreign nationals. Then France proposed a conference of foreign Ambassadors in Constantinople to concert measures for suppressing the Arabi revolt. Before it had time to meet, a riot broke out in Alexandria, in which more than fifty Europeans were killed. At the end of June 1882 the Ambassadors signed a protocol, agreeing that if intervention became necessary their respective countries would not seek territorial advantage or an exclusive position in Egypt. Meanwhile Arabi prepared to resist a foreign invasion and started to strengthen the fortifications of Alexandria. When he refused to stop this work, the British Admiral in command, acting on instructions from London, opened fire and destroyed his forts. British troops were then landed, acting in the name of the Khedive, to quell the disorders which followed. The French squadron meanwhile had sailed away to Port Said before the start of the bombardment.

Britain now turned her attention to the protection of the Suez Canal, appealing for military support from the powers and particularly from France. She received authority from the Khedive to occupy Port Said, Suez and various points on the Canal, thus ensuring freedom of transit and protecting the population of the Isthmus. Though a Liberal Government, under Gladstone, was now back in power, the necessary credits were voted and Parliament sanctioned the intervention.

France, at this juncture, had a weak, vacillating Government, divided and unable to control its Chamber, which thus refused to grant the necessary credits—largely owing to strong opposition from Clemenceau. This failure to intervene cost France her long-standing position of influence in Egypt. It arrested a process of history and left the field open to Britain, who now intervened alone.

At this point Lesseps, now seventy-seven years old, arrived in Egypt, passionately concerned to uphold the 'absolute neutrality' of his Canal, as laid down in the Viceroy's Concession and the Sultan's firman. From Arabi, whose Nationalist movement he favoured, he claimed to have received an undertaking to respect this neutrality, provided the British forces did so too, refraining from the disembarkation of troops at Port Said, Suez and elsewhere. If they disembarked, on the other hand, Arabi—so Lesseps insisted—had plans to block the Sweet Water Canal, and thus deprive the Isthmus of water. For three weeks Lesseps agitated, hurrying from one end of his Canal to the other, prodding, preparing, arguing, warning, and once threatening the British Admiral that he would personally blow out the brains of the first British officer who dared to land. 'The old rogue is playing his tricks', observed Granville, once again Foreign Minister. 'I trust we shall get the better of him.'

All his protests were of no avail. On the night of August 19, 1882, a British expeditionary force under General Sir Garnet Wolseley landed at Port Said, while further British forces proceeded down the Canal to occupy Ismailia. Meanwhile reinforcements arrived from India at Suez. The operation, it was claimed, was defensive, carried out in the name and for the protection of the legitimate ruler of Egypt. No retaliatory move was made against the Canal by Arabi who, after hesitating, decided temporarily to block it, but acted too late. As soon as the British forces were ready for action they marched against him and defeated his armies in a single morning at Tel-el-Kebir, on the fringe of the Delta. Next day they occupied Cairo.

The Dual Control was no more, and Britain resisted a move by France to revive it. Egypt was now hers alone. Earlier in the century British forces had twice landed in Egypt—once successfully in 1801, to eject Napoleon, and once, unsuccessfully in 1807, to dethrone Mohammed Ali. But each time they withdrew. For it was then no part of British policy to seek such territorial commitments. Now, thanks to the opening of the Suez Canal, the situation was different. Gladstone might insist that the occupation was no more than temporary; otherwise it would be 'absolutely at variance with all the principles and views of Her Majesty's Government, and the pledges they have given to Europe'; and his successors were to repeat this insistence for many years to come. There could be no question of annexation, since Egypt was still part of the Ottoman Empire. Instead Britain established, in effect, a form of protectorate over the country, hence over the Canal. Her occupation was to last, in one form or another, for seventy-four years.

Meanwhile the obstructive tactics of Lesseps, and their implications for the future, had agitated opinion in Britain and disquieted its Government. This led to the revival of a scheme, discussed ten years earlier, to build in competition a second Canal, under the control of a British Company, from Alexandria across the Delta to Suez. Lesseps objected to the plan as an infringement of a monopoly, granted in his original Concession, for canal construction between the two seas. His claim was upheld. But to conciliate the British Government, he agreed that his own Company would construct this second Canal, subject to British pressure on the Egyptian Government for the necessary Concession and lands, and to a British loan not exceeding £8 million. This measure however was decisively defeated in Parliament. Instead the Company came to amicable terms with Lord Granville on increased British representation and co-operation. They agreed on a future programme involving such matters as the reduction of tolls in relation to profits, and the enlargement and improvement of the Canal in relation to increases in traffic.

It now only remained to negotiate the neutralization of the Canal. Lesseps had been at fault—as indeed he well knew—in insisting on the fact of its neutrality. He was arguing on the basis of a principle proclaimed unilaterally by Egypt and Turkey, but not guaranteed internationally by the European powers concerned. Earlier in the century, when a Canal across the Isthmus of Suez existed only as a nebulous project in the minds of men, it was envisaged as a neutral waterway, on the lines of the Bosporus, by Metternich, who put forward the idea, without success, to Mohammed Ali. In 1856, when it existed on paper, Lesseps, on Austrian advice, put forward a plan for its neutralization to the Congress of Paris. But he withdrew it on account of opposition from Britain. In 1864 he put forward another plan to Drouyn de Lhuys, and after the opening of the Canal, in 1869, the International Commercial Congress, in Cairo, voted for an international guarantee of its neutrality.

Now, in 1883, Lord Granville proposed to the powers a general agreement for neutralization; and France, to achieve it, made a specific proposal for an International Commission. Britain however was not in a hurry to see her own anomalous position in 'temporary' occupation of Egypt called thus openly in question— an occupation which, as it was, promised her effective control of the Canal in the event of war. But eventually, in October 1888, a Convention was signed at Constantinople by the representatives of nine powers, which, allowing for British and French reservations, established 'a definite system destined to guarantee at all times and for all Powers the free use of the Suez Maritime Canal'. It was to be open to all vessels, in time of war as of peace; its entrances were not to be blockaded; no permanent fortifications were to be erected on its banks; no belligerent warships must disembark troops or munitions in its ports or within it. If Egypt were unable to defend it, she could appeal to Turkey, or through Turkey to the signatory powers. So the course of the new Canal, now secure in its rights of international passage, as already in its internal system of operation, ran smoothly ahead towards a distant future.[1]

*   *   *

In 1876 a writer in the *Quarterly Review*[2] foretold fancifully: 'As in Egypt of old a Pharoah arose who knew not Joseph, so, ere long, a generation will come which forgets Lesseps.' The Suez

[1] Ferdinand de Lesseps died in 1894. In 1879, at the age of seventy-four, he had undertaken, on behalf of French interests and in response to strong pressure from public opinion, the construction of the Panama Canal. This time the complexities of politics and finance proved his undoing. A scandal arose, which led, in 1888, to the winding-up of the Panama Company, and the prosecution of Lesseps, his son Charles, and their associates. Charles de Lesseps succeeded in shielding his father by bearing the brunt of the charges himself, and spent three years in 'exile' in Britain, where his name and connection with the Suez Canal won him numerous friends.   [2] October 1876.

Canal was due, under the original Concession of 1854 and the Sultan's firman, to revert to the Egyptian Government in 1968—ninety-nine years after its opening. The generations passed until, in 1952, the Egyptian Government became a Republic, ruled by just such a military régime as Arabi had briefly established, with Gamal Abdel Nasser, like Arabi a fellah by origin, at the head of it. It brought to an end the dynasty of Mohammed Ali, driving into exile his great-great-grandson, King Farouk. In 1954 it brought to an end the British occupation of Egypt under a treaty by which Britain withdrew her forces from the zone of the Canal. In 1956, a month after the last British soldier had left, the Egyptian Government seized and nationalized the Suez Canal, which thus, twelve years before the scheduled date, came under exclusive Egyptian control.

To counter this unilateral action Britain, with the other maritime powers, sought to restore, by negotiation, the Canal's international status, as laid down in the Convention of 1888. She failed to do so, and in concert with France and the new state of Israel, sent a military force into Egypt to reoccupy the Canal. Troops were landed at Port Said, under cover of a naval bombardment, and the Egyptian garrison surrendered. But in response to international pressure, a cease-fire was called, and the Anglo-French forces withdrew before the Canal could be taken. By this intervention, Britain effectively lost all influence in Egypt, as France had lost it by her earlier non-intervention.

Egyptian troops reoccupied Port Said, and threw down the statue of Lesseps which commanded its harbour, leaving only a large empty pedestal for the new generation which was to know him not. Later, a booklet was issued by the Egyptian Suez Canal Authority, in which no mention was made of him, only of the hundreds of thousands of Egyptians who had laboured in the hot sun for ten long years to create the Canal.

Ten years thus elapsed. Then, in June 1967, the forces of Israel invaded Egypt once more, and occupied the Asiatic bank of the Suez Canal. It was now indeed the 'fortified ditch' of Lord Palmerston's fears. But France, Britain and the Ottoman Empire were forgotten, and it was the returning children of Joseph and

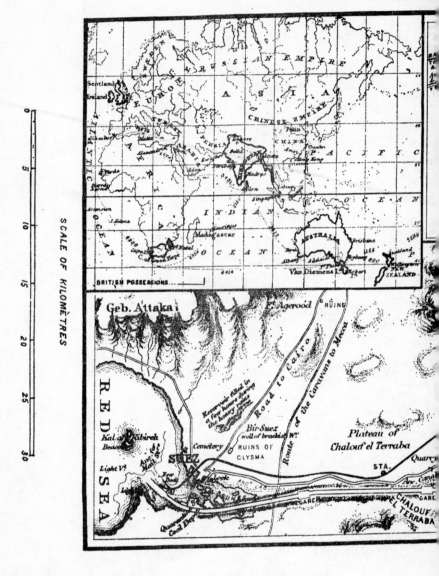

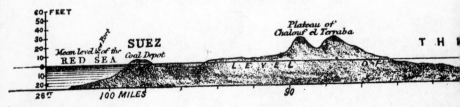

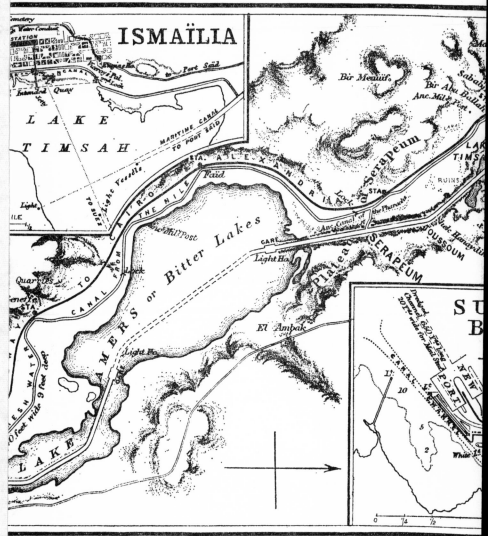

ISMAÏLIA

SECTION OF THE ISTHMU

LONDON, PUBLISHED BY JAMES WYLD, GEOGRAPHER TO THE QUEE

the children of Pharaoh who faced each other across its waters. So, in its ninety-ninth year the Suez Canal was, for the first time, indefinitely closed to all traffic; and, as before its existence, the ships of the world sailed again around the Cape of Good Hope.

London 1968

Select Bibliography
Index

Select Bibliography

PRIMARY SOURCES

Foreign Office Records, Series No. 78 (Prefix F.O.)
Archives Françaises, Correspondance Politique, Egypte et Turquie (Prefix C.P.);
Mémoires et Documents Egypte (Prefix M & D)
Bulwer Papers
Lesseps, Ferdinand de: *Lettres, Journal et Documents, pour servir à l'Histoire du Canal de Suez*. 5 Vols. 1854–1869. Paris 1881. (Abbreviation *Journal*)

SECONDARY SOURCES
General

Batbedat, Th.: *De Lesseps Intime,* Paris 1913
Beatty, Charles: *De Lesseps of Suez,* London 1956
Charles-Roux, J.: *L'Isthme et le Canal de Suez,* 2 Vols. Paris 1901
Coureau, Robert: *Ferdinand de Lesseps,* Paris 1932
Crabitès, Pierre: *The Spoliation of Suez,* London 1940
Edgar-Bonnet, George: *Ferdinand de Lesseps,* 2 Vols. Paris 1951, 1959.
d'Elbée, Jean: *Un Conquistador de Génie, Ferdinand de Lesseps,* Paris 1943
Hallberg, Charles W.: *The Suez Canal:Its History and Diplomatic Importance.*
  Columbia University, New York 1931
Lesseps, Ferdinand de: *Souvenirs de Quarante Ans,* 2 Vols. Paris 1887
Marlowe, John: *The Making of the Suez Canal,* London 1964
Siegfried, André: *Suez and Panama,* London 1940
Voisin Bey: *Le Canal de Suez, 1854–1901.* 7 Vols. Paris 1902–7
Wilson, Sir Arnold: *The Suez Canal, Its Past, Present and Future,* London 1933

CHAPTER REFERENCES

I: NAPOLEON AT SUEZ

Baldwin, George: *Political Recollections relative to Egypt,* London 1801
Berthier, Général: *Relations des Campagnes du Général Bonaparte, en Egypt et en Syrie,* Vol. IX. Paris 1801
Bertrand, Général: *Compagnes d'Egypte et de Syrie,* Vol. I, 1798–99. Paris 1847
Browne, Haji A.: *Bonaparte in Egypt,* London 1907

SELECT BIBLIOGRAPHY

Charles-Roux, F.: *Bonaparte: Governor of Egypt*, London 1937
Charles-Roux, J.: *Op. cit.,* Vol. I, Chs. 1–4
Edgar-Bonnet: *Op. cit.,* Vol. I, Pt. 2, Ch. 1
Elgood, Lieut.-Col. P. G.: *The Transit of Egypt*, London 1928
— *Bonaparte's Adventure in Egypt*, Oxford 1931
Hoskins, H. L.: *British Routes to India*, London 1928
Jabarti, Sheikh Abd-el-Rahman el: *Merveilles Biographiques et Historiques ou Chroniques,* Vol. VI. Cairo 1891
Jonquière, C. de la: *L'Expedition d'Egypte (1798–1801),* Vol. III. Paris; 1899–1907.
Napoleon I: *Correspondance,* Vol. V. Paris 1860
Wood, A. C.: *A History of the Levant Company.* Oxford 1935
Description de l'Egypte, ou Receuil des Observations et des Recherches qui ont été faites en Egypte pendant l'Expedition de l'Armée Française, Vol. I, Paris 1809
Memoires sur l'Egypte, publiés pendant les campagnes du Général Bonaparte, Vol. I, 1800–1802. Paris

II: THE OVERLAND ROUTE

Barker, Edward B. B.: *Syria and Egypt under the last five Sultans of Turkey: Experiences during fifty years of Mr. Consul-General Barker,* by his son. London 1870
Bartlett, W. H.: *The Nile Boat, or Glimpses of the Land of Egypt,* London 1850
Cable, Boyd: *A Hundred Year History of the P & O.* London 1937
Charles-Roux, J.: *Op. cit.,* Vol. I, Ch. 5
Dodwell, Henry: *The Founder of Modern Egypt: A Study of Muhammad Ali.* London 1931
Edgar-Bonnet: *Op. cit.,* Vol. I, Ch. 1
Elwood, Mrs. Colonel: *Narrative of a Journey Overland to India.* London 1830
Gorbal, Shafik: *The Beginnings of the Egyptian Question and the Rise of Mehemet Ali.* London 1928
Head, Captain C. F.: *Eastern and Egyptian Scenery, Ruins, etc. . . . illustrative of a journey to Europe, followed by an outline of an overland route, statistical remarks etc., intended to show the advantage and practicability of Steam Navigation from England to India.* London 1833
Pückler-Muskau, Prince: *Egypt under Mehemet Ali.* London 1845
Sidebottom, John K.: *The Overland Mail: A Postal Historical Study of the Mail Route to India.* London 1948
Thackeray, W. M.: *Notes of a Journey from Cornhill to Grand Cairo.* London 1846

SELECT BIBLIOGRAPHY

III: SCHEMES FOR A CANAL

F.O.97/411 (1841–1851)

d'Allemagne, H. R.: *Prosper Enfantin*. Paris 1930

Anderson, Arthur: *Observations on the Practicability and Utility of opening a Communication between the Red Sea and the Mediterranean*. London 1843

Booth, Arthur John: *Saint Simon and Saint Simonism: A Chapter in the History of Socialism in France*. London 1871

Bulwer, Sir H. Lytton: *The Life of Viscount Palmerston, 1784–1846*, Vol. 2. London 1871

du Camp, Maxime: *Souvenirs Littéraires; 1822–1894*. Paris 1962.

Chesney, General F. R.: *Narrative of the Euphrates Expedition*. London 1868

— *Life of* by his wife and daughter, Louisa Chesney and Jane O'Donnell, edited by Stanley Lane-Poole. London 1885

Edgar-Bonnet: *Op. cit.*, Vol. I, Ch. 1

Hallberg: *Op. cit.*, Chs. 5, 6, 7, 8, 9

Stephenson, Robert: *Isthmus of Suez Canal 1858* (Letter to editor of *Austrian Gazette*)

— *Life of*, by J. C. Jeaffreson. London 1864

House of Commons: Select Committee on Steam Navigation to India. 1834.

IV: A CONCESSION FOR LESSEPS

Journal. Vol. I, 1854–1855

Bridier, L.: *Une Famille Française: Les de Lesseps*. Paris 1900

Charles-Roux, J.: *Op. cit.*, Vol. I, Ch. 7

Edgar-Bonnet: *Op. cit.*, Vol. I, Pt. 2, Chs. 4, 5; Pt. 2, Chs. 2, 3

Hallberg: *Op. cit.*, Ch. 10

V: OBSTACLES AT CONSTANTINOPLE

F.O. 78/1156 (1854–1855)

C.P. Egypt, 25, 26 (1853–1856)

Journal: Vol. I (1855)

Charles-Roux, J.: *Op. cit.*, Vol. I, Ch. 6

Edgar-Bonnet: *Op. cit.*, Vol. I, Pt. 1, Ch. 5

Hallberg: *Op. cit.*, Ch. 10

Lane-Poole, Stanley: *Life of Lord Stratford de Redcliffe*, 2 Vols. London 1888

VI: 'THE MAN OF 1840'

F.O. 78/1156 (1855–1856)

C.P. Egypt, 25, 26 (1854–1856)

Journal. Vol. I (1854–1856)

Ashley, Hon. Evelyn: *The Life of Viscount Palmerston,* 2 Vols. London 1846–1865

Bell, H. C. F.: *Lord Palmerston,* 2 Vols. London 1936

Charles-Roux, J.: *Op. cit.,* Vol. I, Ch. 6

Edgar-Bonnet: *Op. cit.,* Vol. I, Pt. 2, Ch. 5

Hallberg: *Op. cit.,* Ch. 10

Guedalla, Philip: *Palmerston.* London 1926

— *The Second Empire.* London 1922

Guérard, Albert: *Napoleon III.* Harvard 1943

Kurtz, Harold: *The Empress Eugénie.* London 1964

Martin, Kingsley: *The Triumph of Lord Palmerston.* London 1924

Morley, Viscount: *The Life of Richard Cobden.* London 1881

Simpson, F. A.: *Louis Napoleon and the Recovery of France 1848–1856.* London 1923

Taylor, A. J. P.: *The Struggle for Mastery in Europe, 1848–1918.* Oxford 1954

Wellesley, Colonel Hon. F. A.: *The Paris Embassy during the Second Empire, 1852–1867.* London 1925

Economist, Athenaeum, Edinburgh Review (1855–1856)

VII: TOWARDS THE *FAIT ACCOMPLI*

F.O. 78/1340 (1856–1857); 1421 (1858)

C.P. Egypt, 26 (1855–1856); Turkey, 333 (1857)

M & D XIII (1855)

Journal: Vol. I (1856), Vol. II (1857–1858), Vol. V, Appendix (1870)

Bell: *Op. cit.*

Charles-Roux, J.: *Op. cit.,* Vol. I, Ch. 6; Annexes 14, 15, 16

Edgar-Bonnet: *Op. cit.,* Vol. I, Pt. 2, Chs. 6, 7

Hallberg: *Op. cit.,* Ch. 11

Hansard's Parliamentary Debates, 1857, 1858, 1861

Daily News, Spectator, Bristol Advertiser, Commercial Record, Mercantile Journal (1857)

VIII: FORMATION OF THE COMPANY

F.O. 78/1421 (1858); 1489 (1859)

C.P. Egypt 28 (1858–1859); Turkey, 334, 335

M & D XIII (1857)

Journal, Vol. II (1857–1858)

Charles-Roux, J.: *Op. cit.,* Vol. I, Ch. 7

Edgar-Bonnet: *Op. cit.*, Vol. I, Pt. 2, Chs. 7, 8
Hallberg: *Op. cit.*, Ch. 11, 12
The Globe (1858)

IX: AN APPEAL TO THE EMPEROR

F.O. 78/1489 (1859); 1556 (1860)
C.P. Egypt 28 (1859); Turkey, 341, 342 (1859)
M & D XIII (1859)
Journal: Vol. III (1859–1860)
Charles-Roux, J.: *Op. cit.*, Vol. I, Chs. 7, 8
Edgar-Bonnet: *Op. cit.*, Vol. I, Pt. 2, Ch. 8
Hallberg: *Op. cit.*, Ch. 12
The Times, Morning Herald, Daily News, Morning Post (1859)

X: A CANAL TAKES SHAPE

F.O. 78/1556 (1860); 1715 (1861–1862); 1795 (1863)
C.P. 29 (1860), 30 (1861–1862), 31 (1863)
Bulwer Papers (1862)
Journal: Vol. 3 (1866), Vol. 4 (1861, 1862, 1863)
Charles-Roux, J.: *Op. cit.*, Vol. I, Pt. 1, Chs. 8, 11
Edgar-Bonnet: *Op. cit.*, Vol. I, Part 2, Ch. 8
Fitzgerald, P.: *The Great Canal at Suez*, Vol. I. London 1876
Hallberg: *Op. cit.*, Ch. 12
Landes, David S.: *Bankers and Pashas*, Ch. 3. London 1958
Hansard's Parliamentary Debates, 1861

XI: ISMAIL TAKES OVER

F.O. 78/1795, 1796 (1863)
C.P. Egypt 31, 32 (1863); Turkey 357, 359 (1863)
Bulwer Papers (1863)
Journal: Vol. 4 (1863)
Charles-Roux, J.: *Op. cit.*, Vol. I, Ch. 8
Douin, G.: *Histoire du Règne du Khedive Ismail*, Vol. I, Chs. 1, 2. Rome 1933
Edgar-Bonnet: *Op. cit.*, Vol. I, Pt. 2, Chs. 8, 9
Hallberg: *Op. cit.*, Chs. 12, 13
Landes: *Op. cit.*, Ch. 7.
Moberly Bell, C. F.: *Khedives and Pashas*. London 1884
Sabry, M.: *L'Empire Egyptien sous Ismail et L'Ingerence Anglo-Française (1863–1879)*, Ch. 7. Paris 1933
Hansard's Parliamentary Debates, 1862
The Spectator (1863)

SELECT BIBLIOGRAPHY

XII: INTRIGUES IN PARIS

F.O. 78/1795, 1796 (1863), 1849 (1864)
C.P. Egypt 31, 32 (1863); Turkey, 357, 359 (1863)
Bulwer Papers (1863)
Journal: Vol. IV (1863, 1864)
Charles-Roux, J.: *Op. cit.,* Vol. I, Ch. 8; Annexe 18
Douin: *Op. cit.,* Vol. I, Chs. 2, 3
Edgar-Bonnet: *Op. cit.,* Vol. I, Pt. 2, Ch. 9
Guedalla: *The Second Empire, Op. cit.,* Pts. III, IV
Hallberg: *Op. cit.,* Ch. 13
Kurtz: *Op. cit.*
Landes: *Op. cit.,* Ch. 7
Moberly Bell: *Op. cit.*

XIII: THE IMPERIAL ARBITRATION

F.O. 78/1849, 1850 (1864); 1895, 1896, 1897 (1865); 2014 (1866)
C.P. Egypt, 33, 34 (1864), 36, 37 (1865); Turkey, 361, 362, 363 (1864)
M & D XIV (1864–1866)
Journal: Vol. IV (1864), Vol. V (1864, 1865, 1866)
Charles-Roux, J.: *Op. cit.,* Vol. I, Ch. 8, Annexes 19, 20, 21, 22, 23
Douin: *Op. cit.,* Vol. I, Chs. 3, 4, 5, 6, 7
Edgar-Bonnet: *Op. cit.,* Vol. I, Pt. 2, Ch. 9, 10
Hallberg: *Op. cit.,* Ch. 13
Landes: *Op. cit.,* Ch. 7
Sabry: *Op. cit.,* Ch. 7

XIV: THE CANAL IS COMPLETED

Charles-Roux, J.: *Op. cit.,* Vol. I, Ch. 8
Douin: *Op. cit.,* Vol. II, Ch. 4, 16
Edgar-Bonnet: *Op. cit.,* Vol. I, Pt. 2, Ch. 11
Fitzgerald: *Op. cit.,* Vol. I, Ch. 10
Hallberg: *Op. cit.,* Ch. 13, 14
Lee, Sidney: *King Edward VII,* Ch. 7. London 1925
Sabry: *Op. cit.,* Ch. 7
Fortnightly Review, Captain J. Clerk on *Suez Canal.* January 1869
The Times, Illustrated London News (1869)

SELECT BIBLIOGRAPHY

XV: THE OPENING CEREMONIES

F.O. 78/2118 (1868–1869)
Journal: Vol. V (1869)
Carré, Jean-Marie: *Voyageurs et Ecrivains Français en Egypte,* Vol. II. Cairo 1956
Charles-Roux, J.: *Op. cit.,* Ch. 9
Douin: *Op. cit.,* Vol. II, Ch. 16
Dicey, Edward: *The Story of the Khedivate.* London 1902
Edgar-Bonnet: *Op. cit.,* Vol. I, Pt. 2, Ch. 12
Mariette Pasha, F. A. F.: *Oeuvres Diverses,* with biographical notice by G. Maspero. London 1904
The Times, Pall Mall Gazette, Blackwood's Magazine (1869–1870)

XVI: 'YOU HAVE IT, MADAM!'

F.O. 78/2118 (1869); 2310 (1873); 2432 (1875)
Blake, Robert: *Disraeli,* Ch. 25. London 1967
Buckle, G. E.: *The Life of Benjamin Disraeli,* Vol. V, Ch. 12. London 1920
Charles-Roux, J.: *Op. cit.,* Vol. II, Chs. 10, 12
Douin: *Op. cit.,* Vol. II, Chs. 11–15, 22, 24
Edgar-Bonnet: *Op. cit.,* Vol. II, Pt. 1, Ch. 1, 2
Hallberg: *Op. cit.,* Ch. 14, 15
Landes: *Op. cit.,* Ch. 15
Moberly Bell, C. F.: *Op. cit.*
Sabry: *Op. cit.,* Ch. 5
Hansard's Parliamentary Debates, 1876
The Times, Quarterly Review (1875–1876)

XVII: BRITAIN MOVES IN

F.O. 78/2432 (1875)
Blake: *Op. cit.,* Ch. 25
Charles-Roux, J.: *Op. cit.,* Ch. 11
Cromer, Earl of: *Modern Egypt,* Vol. I. London 1908
Edgar-Bonnet: *Op. cit.,* Vol. II, Pt. 1, Chs. 2, 3, 4
Hallberg: *Op. cit.,* Chs. 16, 17
Monroe, Elizabeth: *Britain's Moment in the Middle East.* London 1964
Quarterly Review (1876)

Index

Serapeum, 16, 149, 232, 235, 251
Shepheard's Hotel, Cairo, 239–40
Sherif Pasha, 125, 131, 229, 266, 268
Sinai, 3, 34, 65–6, 253
Sixtus V, Pope, 8
Société d'Etudes du Canal de Suez, 49
Société Générale, 263, 266–7
Spain, 55–6, 138; attitude to Canal, 125, 136
Stanton, Colonel (later Major-General) Sir Edward, 211, 227; and British purchase of Khedive's shares, 260, 264–6, 268, 274–5
Steam Committee, Indian, 19, 21, 24
Stephenson, Robert, 50–1, 92, 107–8, 111
Stokes, Lieut.-General Sir John, 272
Stratford de Redcliffe, Viscount, 68–76, 101–2, 108
Sudan, 104
Suez, Napoleon at, 1–4; harbour, 2, 17, 93, 196, 221, 231; British ships at, 11–13; 'Frankish' ships forbidden access to, 12; regular route between Alexandria and, 28–32; railway between Alexandria, Cairo and, 42, 47–8, 51–3, 68, 172; Lesseps and engineers visit, 65–6; quarries at, 122, 150; service canal to, 151; Sweet Water Canal to, 152, 194, 221; a French town, 156; Anglo-French shipping dispute over, 211; coal to, *via* Canal, 221, 247; Prince and Princess of Wales at, 233; opening ceremonies at, 251–3; British occupation of, 278–9; canal suggested between Alexandria and, 280
Suez, Gulf of, 5, 6, 12, 50
Suez, Isthmus of, once a strait, 4; Egyptian frontier, 6; French plans for trade route over, 9–11; lacks stone quarries, 150; Said visits, 152–3; growth of towns and workshops on, 215; Chamber of Commerce delegates visit, 218–221; a free zone, open to contraband, 229
Surveys of: Le Père's, 4, 14–17, 50, 62; Chesney's, 34–5; Linant Bey's, 36–7, 50; The Study Group's, 50–2, 62; Enfantin's, 42; Gallice

and Mougel's, 62; Linant Bey and Mougel's, 65–6; International Scientific Commission's, 92–3; Lesseps's, before starting construction, 120–2
Suez Canal:
Precursors: Canal of Pharoahs (seventh cent. B.C.), 1, 4–5, 34, 219; Persians (sixth cent. B.C.), 5, 17; Ptolemies (third cent. B.C.), 5; Romans (second cent. B.C.), 5; Caliphs (seventh cent.), 5–6, 10
Plans for revival of: sixteenth cent., 7–8; seventeenth cent., 8–10; eighteenth cent. (French) 10, (British) 11–13, (Napoleon's) 2, 4, 6, 14, 17, 62; nineteenth cent. (French) 2, 4, 6, 14, 39–40, 42, 48, 62, (British) 34–8, (German) 49
Reports on: Le Père's, 15–17, 41, 62; Chesney's, 35–6; Linant Bey's, 37, 62; Talabot's, 50–2, 62, 65; Gallice and Mougel's, 62; Linant Bey and Mougel's, 66, 93; International Scientific Commission's, 93, 97; Hawkshaw's, 158, 183, 215, 219, 234; unfavourable, of French engineer, 181
Routing of: indirect, 16–17, 51, 65, 68; direct, 37, 93, 97
Opposition to: British, 48, 51–2, 60, 66–75, 86–92, 94, 99, 105–7, 109–114, 119, 124, 127, 130–1, 163; Abbas Pasha's, 53, 59; French, 88, 113; Turkish, 94–5, 99, 113–14, 124–5
International approach to: 262, 264; of Study Group, 49–52; attempt to hold conference on, 136–41; conference on tolls, 258
Said and: 60–5, 118–19, 144
Tolls: 65, 97; dispute about, 257–259, 262, 273
Neutrality of: Lesseps seeks international guarantee of, 72, 94, 279–81; impracticability of, 91; Metternich on, 100–1, 280; Turkey insists on, 169, 279; appeal to Emperor on, 191; international guarantee of, 281
Porte's consent to: need for, 99, 101,